Certified SOLIDWORKS Professional
Advanced Preparation Materials
SOLIDWORKS 2018

SDC
Publications

SDC Publications
P.O. Box 1334
Mission, KS 66222
913-262-2664
www.SDCpublications.com
Publisher: Stephen Schroff

Acknowledgments

Thanks as always to my wife Vivian and my daughter Lani for always being there and providing support and honest feedback on all the chapters in the textbook.

I would like to give a special thanks to Rachel Schroff for her editing and corrections. Additionally, thanks to Kevin Douglas and Peter Douglas for writing the forewords.

I also have to thank SDC Publications and the staff for its continuing encouragement and support for this edition of **SOLIDWORKS 2018 CSWPA Preparation Materials**. Thanks also to Gretchen Hughes for putting together such a beautiful cover design.

Finally, I would like to thank you, our readers, for your continued support. It is with your consistent feedback that we were able to create the lessons and exercises in this book with more detailed and useful information.

Foreword

For more than two decades, I have been fortunate to have worked in the fast-paced, highly dynamic world of mechanical product development providing computer-aided design and manufacturing solutions to thousands of designers, engineers and manufacturing experts in the western US. The organization where I began this career was US CAD in Orange County, CA, one of the most successful SOLIDWORKS Resellers in the world. My first several years were spent in the sales organization prior to moving into middle management and ultimately President of the firm. In the mid-1990s is when I met Paul Tran, a young, enthusiastic instructor who had just joined our team.

Paul began teaching SOLIDWORKS to engineers and designers of medical devices, automotive and aerospace products, high tech electronics, consumer goods, complex machinery and more. After a few months of watching him teach and interacting with students during and after class, it was becoming pretty clear – Paul not only loved to teach, but his students were the most excited with their learning experience than I could ever recall from previous years in the business. As the years began to pass and thousands of students had cycled through Paul's courses, what was eye opening was Paul's continued passion to educate as if it were his first class and students in every class, without exception, loved the course.

Great teachers not only love their subject, but they love to share that joy with students – this is what separates Paul from others in the world of SOLIDWORKS Instruction. He always has gone well beyond learning the picks & clicks of using the software, to best practice approaches to creating intelligent, innovative and efficient designs that are easily grasped by his students. This effective approach to teaching SOLIDWORKS has translated directly into Paul's many published books on the subject. His latest effort with SOLIDWORKS 2018 is no different. Students that apply the practical lessons from basics to advanced concepts will not only learn how to apply SOLIDWORKS to real world design challenges more quickly, but will gain a competitive edge over others that have followed more traditional approaches to learning this type of technology.

As the pressure continues to rise on U.S. workers and their organizations to remain competitive in the global economy, raising not only education levels but technical skills is paramount to a successful professional career and business. Investing in a learning process towards the mastery of SOLIDWORKS through the tutelage of the most accomplished and decorated educator and author in Paul Tran will provide a crucial competitive edge in this dynamic market space.

Kevin Douglas
Vice President Sales/Board of Advisors, GoEngineer

Preface

I first met Paul Tran when I was busy creating another challenge in my life. I needed to take a vision from one man's mind, understand what the vision looked like, how it was going to work and comprehend the scale of his idea. My challenge was I was missing one very important ingredient, a tool that would create a picture with all the moving parts.

Research led me to discover a great tool, SOLIDWORKS. It claimed to allow one to make 3D components, in picture quality, on a computer, add in all moving parts, assemble it and make it run, all before money was spent on bending steel and buying parts that may not fit together. I needed to design and build a product with thousands of parts, make them all fit and work in harmony with tight tolerances. The possible cost implications of failed experimentation were daunting.

To my good fortune, one company's marketing strategy of selling a product without an instruction manual and requiring one to attend an instructional class to get it, led me to meet a communicator who made it all seem so simple.

Paul Tran has worked with and taught SOLIDWORKS as his profession for 30 years. Paul knows the SOLIDWORKS product and manipulates it like a fine musical instrument. I watched Paul explain the unexplainable to baffled students with great skill and clarity. He taught me how to navigate the intricacies of the product so that I could use it as a communication tool with skilled engineers. ***He teaches the teachers***.

I hired Paul as a design engineering consultant to create the thousands of parts for my company's product. Paul Tran's knowledge and teaching skill has added immeasurable value to my company. When I read through the pages of these manuals, I now have an "instant replay" of his communication skill with the clarity of having him looking over my shoulder - ***continuously***. We can now design, prove and build our product and know it will always work and not fail. Most important of all, Paul Tran helped me turn a blind man's vision into reality and a monument to his dream.

Thanks Paul.

These books will make dreams come true and help visionaries change the world.

Peter J. Douglas
CEO, Cake Energy, LLC

Images courtesy of C.A.K.E. Energy Corp., designed by Paul Tran

Author's Note

CSWP - Advanced Preparation Materials are comprised of lessons based on the feedbacks from Paul's former CSWPs students and engineering professionals. Paul has 32 years of experience in the fields of mechanical and manufacturing engineering; 22 years were in teaching and supporting the SOLIDWORKS software and its add-ins. As an active Sr. SOLIDWORKS instructor and design engineer, Paul has worked and consulted with hundreds of reputable companies including IBM, Intel, NASA, US-Navy, Boeing, Disneyland, Medtronic, Oakley, Kingston, Community Colleges, Universities, and many others. Today, he has trained nearly 10,000 engineering professionals, and given guidance to half of the number of Certified SOLIDWORKS Professionals and Certified SOLIDWORKS Experts (CSWP & CSWE) in the state of California.

Every lesson in this book was created based on the actual CSWPA Examinations. Each of these projects have been broken down and developed into easy and comprehendible steps for the reader. Furthermore, every challenge is explained very clearly in short chapters, ranging from 30 to 50 pages. Each and every single step comes with the exact screen shot to help you understand the main concept of each design more easily. Learn the CSWP Advanced Preparation Materials at your own pace, as you progress from Parts, Assemblies, Drawings and then to more complex design challenges.

About the Training Files

The files for this textbook are available for download on the publisher's website at www.SDCpublications.com/downloads/978-1-63057-144-3. They are organized by the chapter numbers and the file names that are normally mentioned at the beginning of each chapter or exercise. In the **Completed Parts** folder you will also find copies of the parts, assemblies and drawings that were created for cross referencing or reviewing purposes.

It would be best to make a copy of the content to your local hard drive and work from these documents; you can always go back to the original training files location at anytime in the future, if needed.

Who this book is for

This book is for the mid-level to advanced user, who is already familiar with the SOLIDWORKS program. To get the most out of this CSWPA-Certification Preparation book it is strongly recommended that you have studied and completed all the lessons in the Basic and Advance textbooks. It is also a great resource for the more CAD literate individuals who want to expand their knowledge of the different features that SOLIDWORKS 2018 has to offer.

The organization of the book

The chapters in this book are organized in the logical order in which you would learn the advanced SOLIDWORKS 2018 topics. Each chapter will guide you through some different tasks, from designing or repairing a mold, to developing a complex sheet metal part; from some

3D sketch for weldments to advancing through more complex tasks that are common to all surface modeling challenges. You will also learn to work with part and assembly drawings and managing different levels of the Bill of Materials.

The conventions in this book

This book uses the following conventions to describe the actions you perform when using the keyboard and mouse to work in SOLIDWORKS 2018:

Click: means to press and release the mouse button. A click of a mouse button is used to select a command or an item on the screen.

Double Click: means to quickly press and release the left mouse button twice. A double mouse click is used to open a program or show the dimensions of a feature.

Right Click: means to press and release the right mouse button. A right mouse click is used to display a list of commands, a list of shortcuts that is related to the selected item.

Click and Drag: means to position the mouse cursor over an item on the screen and then press and hold down the left mouse button; still holding down the left button, move the mouse to the new destination and release the mouse button. Drag and drop makes it easy to move things around within a SOLIDWORKS document.

Bolded words: indicated the action items that you need to perform.

Italic words: Side notes and tips that give you additional information, or to explain special conditions that may occur during the course of the task.

Numbered Steps: indicates that you should follow these steps in order to successfully perform the task.

Icons: indicates the buttons or commands that you need to press.

SOLIDWORKS 2018

SOLIDWORKS 2018 is program suite, or a collection of engineering programs that can help you design better products faster. SOLIDWORKS 2018 contains different combinations of programs; some of the programs used in this book may not be available in your suites.

Start and exit SOLIDWORKS

SOLIDWORKS allows you to start its program in several ways. You can either double click on its shortcut icon on the desktop, or go to the Start menu and select the following: All Programs / SOLIDWORKS 2018 / SOLIDWORKS, or drag a SOLIDWORKS document and drop it on the SOLIDWORKS shortcut icon.

Before exiting SOLIDWORKS, be sure to save any open documents, and then click File / Exit; you can also click the X button on the top right of your screen to exit the program.

Using the Toolbars

You can use toolbars to select commands in SOLIDWORKS rather than using the drop down menus. Using the toolbars is normally faster. The toolbars come with commonly used commands in SOLIDWORKS, but they can be customized to help you work more efficiently.

To access the toolbars, either right click in an empty spot on the top right of your screen or select View / Toolbars.

To customize the toolbars, select Tools / Customize. When the dialog pops up, click on the Commands tab, select a Category, then drag an icon out of the dialog box and drop it on a toolbar that you want to customize. To remove an icon from a toolbar, drag an icon out of the toolbar and drop it into the dialog box.

Using the task pane

The task pane is normally kept on the right side of your screen. It displays various options like SOLIDWORKS resources, Design library, File explorer, Search, View palette, Appearances and Scenes, Custom properties, Built-in libraries, Technical alerts and news, etc.

The task pane provides quick access to any of the mentioned items by offering the drag and drop function to all of its contents. You can see a large preview of a SOLIDWORKS document before opening it. New documents can be saved in the task pane at anytime, and existing documents can also be edited and re-saved. The task pane can be resized, closed or moved to different locations on your screen if needed.

Table of Contents

CSWPA – Drawing Tools

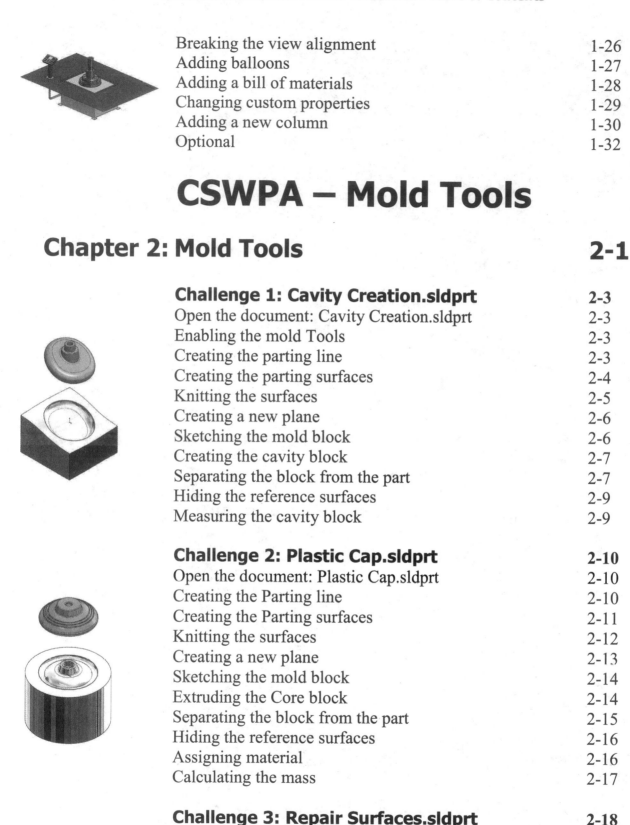

CSWPA – Weldments

Chapter 3: Weldments   3-1

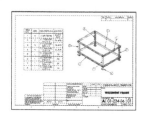

CSWPA – Sheet Metal

Chapter 4: Sheet Metal — 4-1

CSWPA – Surfacing

Chapter 5: Surfacing   5-1

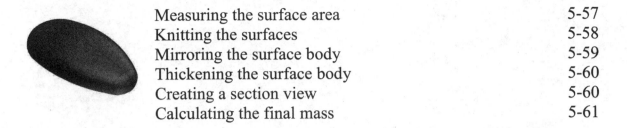

Glossary and Index by chapters

SOLIDWORKS 2018 Quick-Guides:

Quick Reference Guide to SOLIDWORKS 2018 Command Icons and Toolbars.

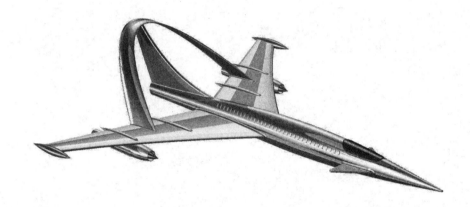

CHAPTER 1

Drawing Tools

CSWPA - Drawing Tools

Certified SOLIDWORKS Professional Advanced Drawing Tools

The completion of the Certified SOLIDWORKS Professional Advanced Drawing Tools (CSWPA-DT) exam proves that you have successfully demonstrated your ability to use the tools found in the SOLIDWORKS Drawing environment.

Employers can be confident that you understand the tools and functionality found in the SOLIDWORKS Drawing environment.

Note: You must use at least SOLIDWORKS 2010 for this exam. Any use of a previous version will result in the inability to open some of the testing files.

Exam Length: 100 minutes

Minimum Passing grade: 75%

Re-test Policy: There is a minimum 30 day waiting period between every attempt of the CSWPA-DT exam. Also, a CSWPA-DT exam credit must be purchased for each exam attempt.

All candidates receive electronic certificates and a personal listing on the CSWP directory when they pass.

Exam features hands-on challenges in many of these areas of SOLIDWORKS drawing functionality such as:

> Basic View Creation, Section Views, Auxiliary Views, Alternate position Views, Broken Out Sections, Lock View/Sheet Focus, Transferring Sketch Entities to/from Views, Bill of materials, and Custom Properties.

CSWPA - Drawing Tools

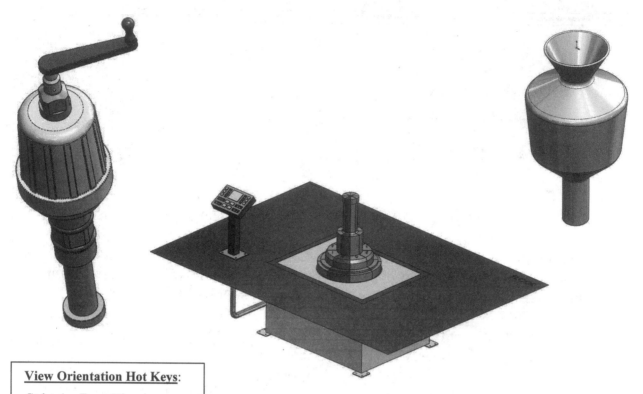

View Orientation Hot Keys:

Ctrl + 1 = Front View
Ctrl + 2 = Back View
Ctrl + 3 = Left View
Ctrl + 4 = Right View
Ctrl + 5 = Top View
Ctrl + 6 = Bottom View
Ctrl + 7 = Isometric View
Ctrl + 8 = Normal To Selection

Dimensioning Standards: **ANSI**

Units: **INCHES** – 3 Decimals

Tools Needed:

 Part Template   Assembly Template   Drawing Template

 View Palette   Section View   Named View

 Measure   Auto Balloon   Bill of Materials

CHALLENGE 1

1. Opening a part document:

- Select **File / Open**.

- Browse to the Training Folder and open the part document named **Tank.sldprt**.

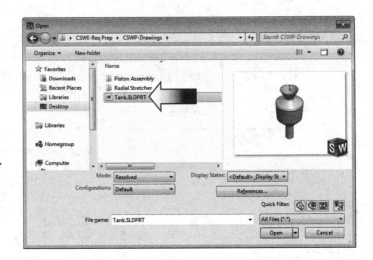

2. Transferring to a drawing:

- Select **File / Make Drawing From Part** (arrow).

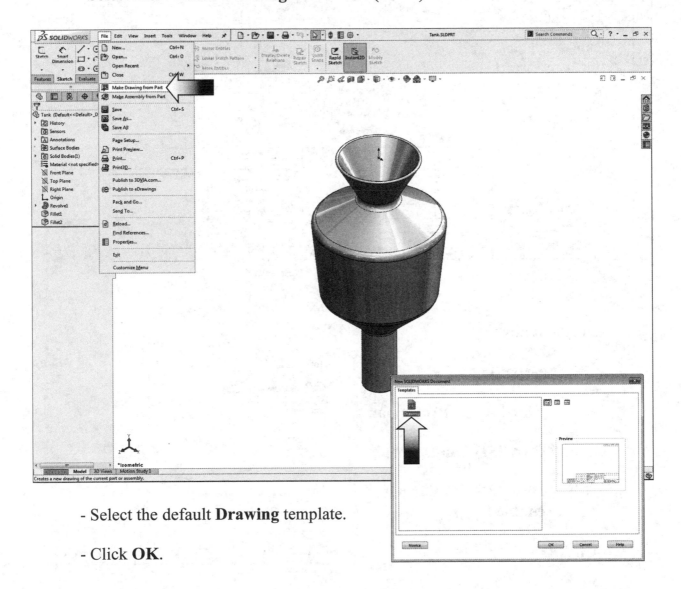

- Select the default **Drawing** template.

- Click **OK**.

3. Changing the paper size:

- Right click inside the drawing and select **Properties**.

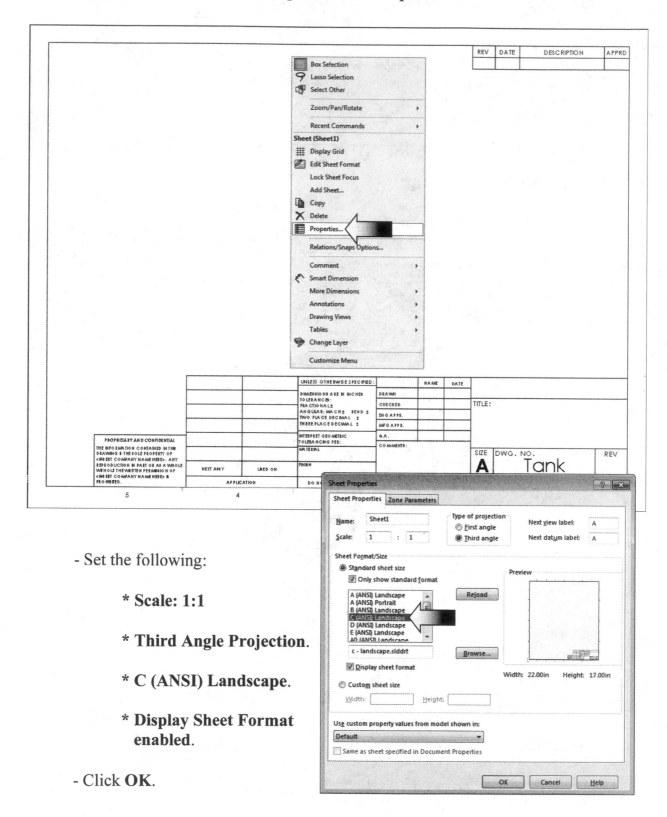

- Set the following:

* **Scale: 1:1**

* **Third Angle Projection**.

* **C (ANSI) Landscape**.

* **Display Sheet Format enabled**.

- Click **OK**.

4. Adding the drawing views:

- Expand the **View Palette** (arrow) and drag the **Front-View** approximately as shown.

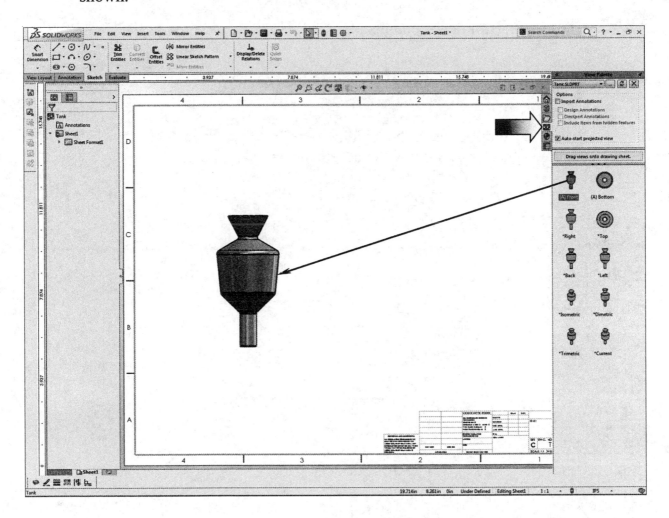

- Project from the Front view or drag and drop the Isometric view from the View Palette.

- Place the Isometric view on the right side of the Front view.

- For clarity, change the tangent edges to With-Font (right click the view's border and select Tangent Edges With Font).

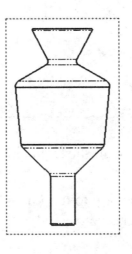

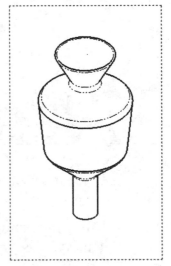

5. Creating a section view:

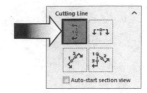

- Change to the **View Layout** tool tab.

- Click the **Section View** command.

- For Cutting Line, select the **Vertical** option (arrow).

- Place the Cutting Line in the middle of the Front view and click the green check mark (arrow) to accept the line placement.

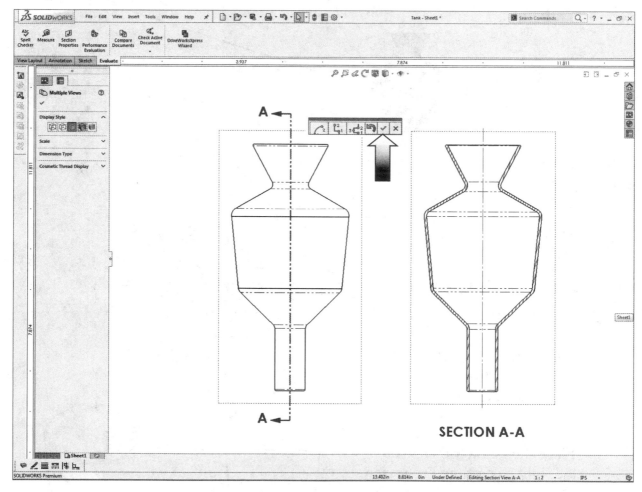

- Place the section view to the right side of the Front view.

- Move the Isometric view to the far right hand side. This view is for reference use only.

6. Measuring the surface area:

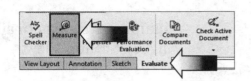

- Zoom in on the section view; we will need to select the sectioned surfaces and measure the total area.

- Change to the **Evaluate** tool tab and select the **Measure** command.

- Hold the Control key and select the 2 sectioned faces as noted.

- Locate the **Total Area** measurement and enter it here:

_____ Inches^2.

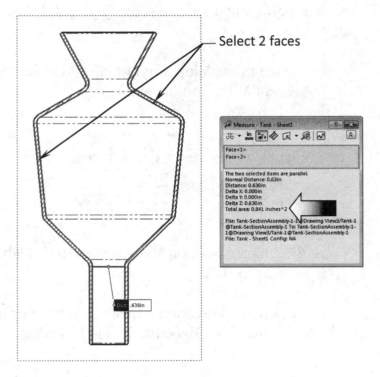

Select 2 faces

7. Creating an aligned section view:

- Double click the dotted border of the Front view to lock it.

- The Lock View Focus option allows you to add sketch entities to a view, even when the pointer is close to another view. You can be sure that the items you are adding belong to the view you want.

- Switch to the **Sketch** tool tab and sketch **2 Lines** as shown.

- Add the vertical and horizontal dimensions to fully define the sketch.

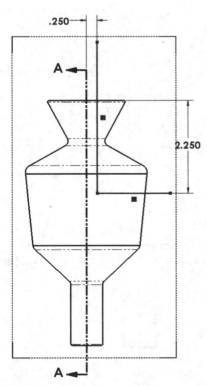

- Multiple lines are normally used to create an Aligned Section View.

- Hold the **Control** key and select the <u>Vertical Line 1st</u>, and then select the <u>Horizontal Line after</u>.

- Switch to the **View Layout** tab and select the **Section View** command (arrow)

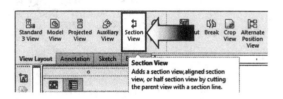

- An Aligned Section View is created and labeled as Section B-B.

- Be sure the Direction Arrows match the image shown below. Click the **Flip Direction** button if needed (arrow).

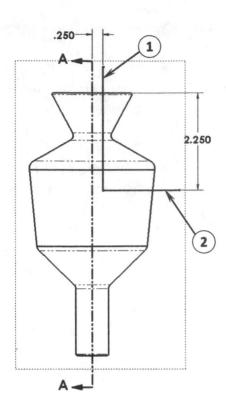

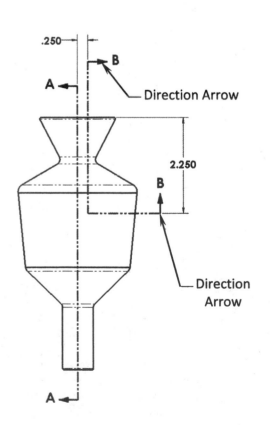

Direction Arrow

Direction Arrow

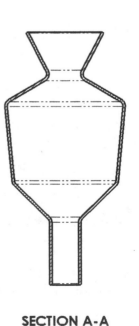

SECTION A-A

SECTION B-B

8. Measuring the surface area:

- Zoom in on the section view; we will need to select the surface of the Section B-B and measure its area.

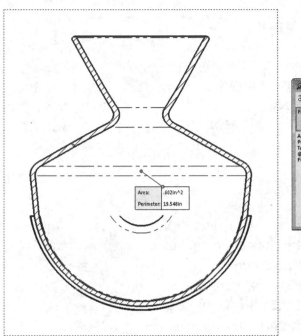

- Change to the **Evaluate** tool tab and select the **Measure** command.

- Select the sectioned face as noted.

- Locate the **Area** measurement and enter it here:

_____ Inches^2.

9. Saving your work:

- Select **File / Save As**.

- Enter **Tank.slddrw** for the file name.

- Click **Save**.

Summary:

The key features to the Challenge 1 are:

- Creating the **Section Views** and **Measuring** the **total surface areas** of the sectioned surfaces.

CHALLENGE 2

1. Opening an assembly document:

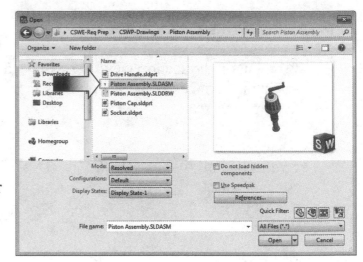

- Select **File / Open**.

- Browse to the Training Folder and open the assembly document named **Piston Assembly.sldasm**.

- In this Challenge, the orientation of the assembly has been changed to some oblique angle.
You will need to come up with a way to find the correct angle and change the orientation of the assembly prior to making a drawing.

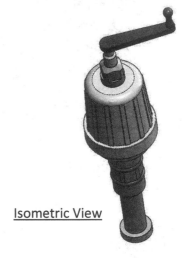

Top View   Isometric View

- Change to the Front orientation (Control+1), the Top orientation (Control+5), the Right orientation (Control+4), and the Isometric view (Control+7) to examine the assembly from different orientations.

- The Top view will be used to correct the orientation of this assembly.

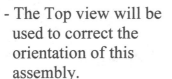

Front View   Right View

2. Creating a reference sketch:

- Open a **new sketch** on the face as indicated.

Sketch face

- We will need to rotate the Handle to the horizontal position. There are several methods to find the current angle of the Handle but we will go with creating a reference sketch approach.

- Sketch a centerline that is **coincident** with the **2 centers** of the crank handle.

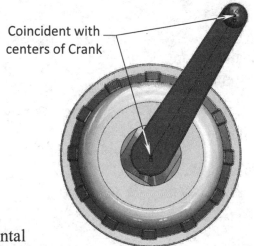

Coincident with centers of Crank

- Sketch a horizontal centerline and **delete** the Horizontal relation, so that we can add the angular dimension without over defining the sketch.

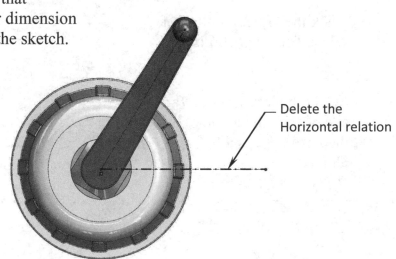

Delete the Horizontal relation

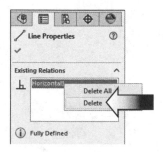

- Add an angular dimension as shown
  to fully define this sketch.

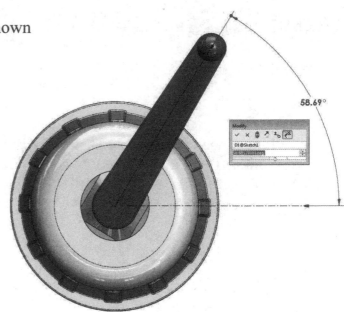

- Keep the default dimension
  value at **58.685°**.

- Highlight the angular dimension and press
  Control+C to copy it to the clipboard.

3. Modifying the view angle:

- Click the **Option** button or
  select **Tools / Options**.

- Select the **View** option (arrow)
  and change the angle of
  the Arrow Keys to
  58.685°.

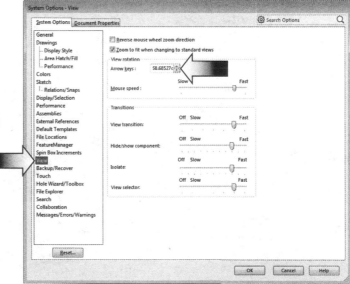

- The upper surface of the Crank Handle should be rotated to a flat position first.

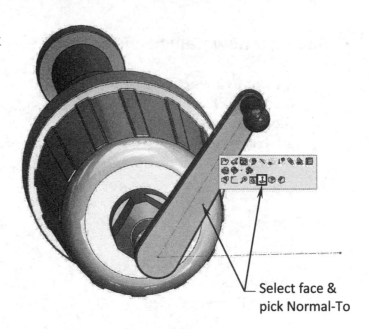

- Click the face as noted and select **Normal-To** (arrow). This option rotates the selected face perpendicular (flat) to the screen.

Select face & pick Normal-To

- Hold the **Alt key** and press the **Left arrow** <u>once</u>, to rotate the view **58.685°** downward.

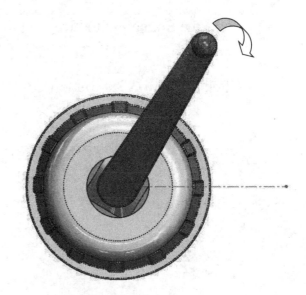

- The Crank Handle and its assembly is now rotated to a horizontal position.

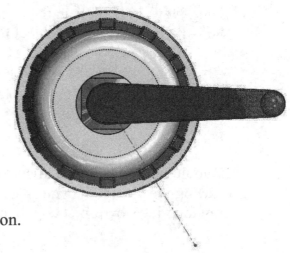

- We will save the new position as a named-view, or a custom view so that we could retrieve it in the drawing later on.

4. Saving a new named-view:

- Custom views can be created and saved in the model or in an assembly so that they can be displayed in a drawing.

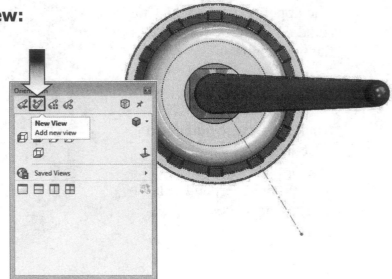

- The views are saved in the Orientation dialog and get carried over to the drawing and listed on the Properties tree.

- Press the **Spacebar** to bring out the Orientation dialog.

- Click the **New View** button 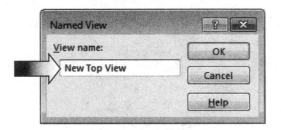.

- Enter: **New Top View** in the Named View dialog and press **OK**.

- The new view is saved and displayed in the Orientation dialog.

- It would be much more difficult to use the original orientations to create the new drawing views in a drawing. The custom view will be used to create the other drawing views by projecting them along the vertical or horizontal directions.

5. Making a drawing from assembly:

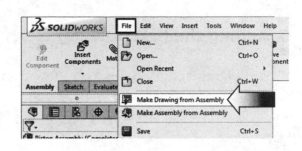

- Select **File / Make Drawing from Assembly** (arrow).

- Select the **Drawing** template.

- The default drawing (A-Size) is displayed. Right click inside the drawing and select **Properties**.

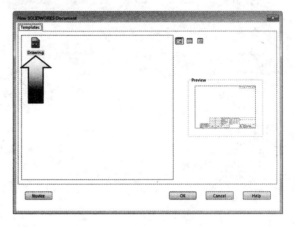

- Change the paper size to **C-Landscape**.

- Change the **Scale** to **1:2**.

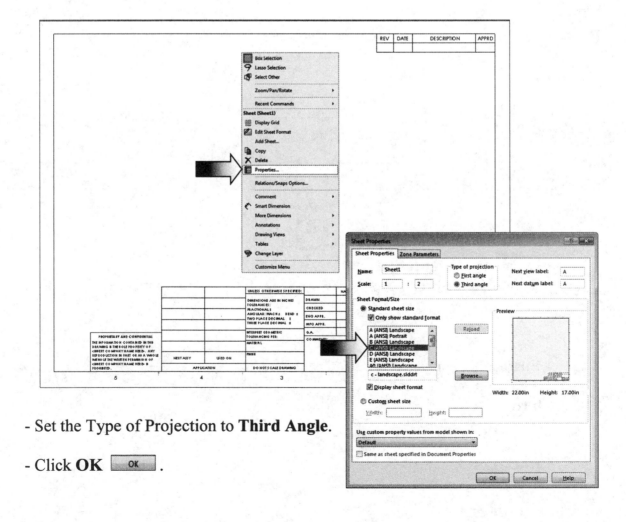

- Set the Type of Projection to **Third Angle**.

- Click **OK** .

6. Adding the first drawing view:

- Drag and drop the **Top** view from the View Palette (change scale to 1:2 if needed).

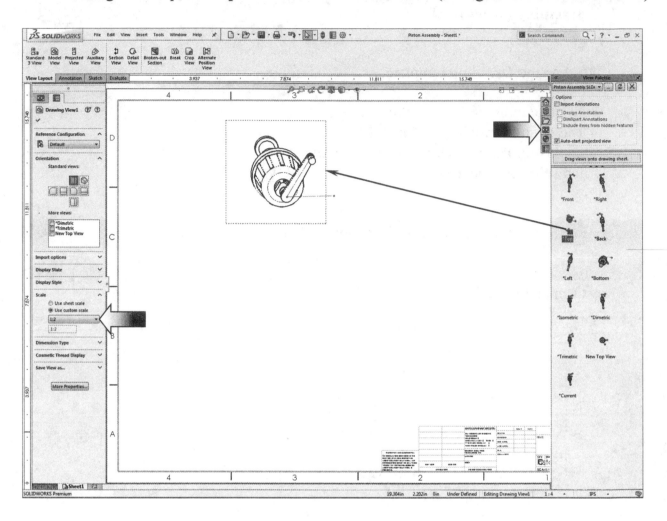

- Locate the **Orientation** section on the Properties tree.

- Enable the named-view **New Top View** check box.

- Click **OK**.

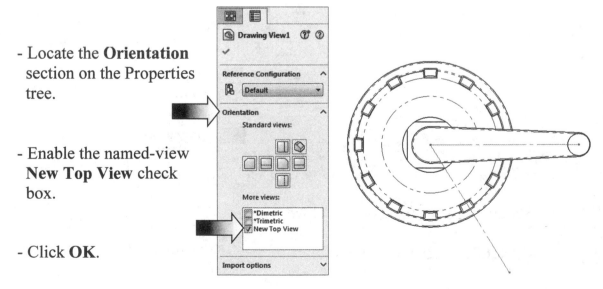

7. Creating the projected drawing views:

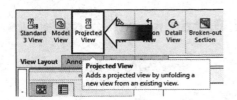

- New drawing views can now be projected vertically or horizontally from the new view.

- Switch to the **View Layout** tool tab and click the **Projected View** command.

- Select the dotted border of the **Top** view to start the projection.

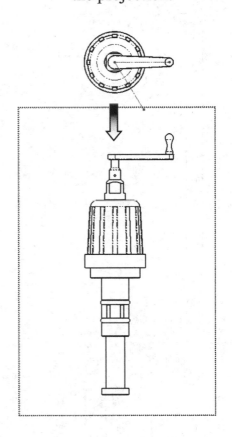

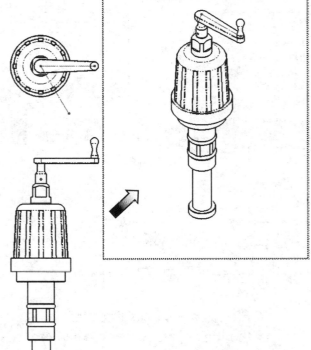

- Move the mouse cursor downward to see the preview of the Front view. Place the Front view under the Top view approximately as shown.

- Additionally, create an isometric view and position it similar to the one shown above.

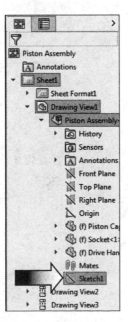

- Locate the **Sketch1** from the Drawing tree and **Show** it .

8. Adding reference lines:

- Zoom in on the lower left corner of the drawing and select the **Line** command from the Sketch tool tab.

- Right click on the dotted border of the Front view and select **Lock View Focus**. This will make the new lines a part of the view. When the drawing view is moved, the lines will also move.

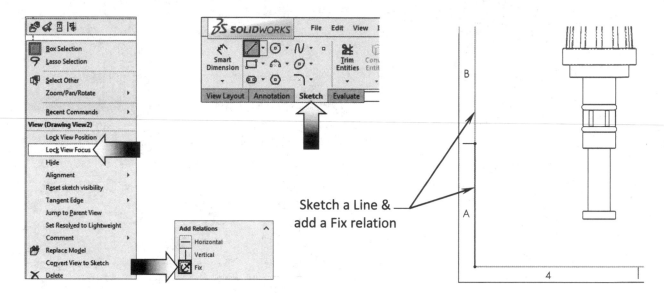

Sketch a Line &
add a Fix relation

- Sketch a **vertical line** starting at the lower left corner of the border.

- Add a **Fix** relation to the line so that it will not move.

- Sketch the **second line** to the right side of the first line approximately as shown.

- Add the **5.00 inches** spacing dimension between the 2 lines.

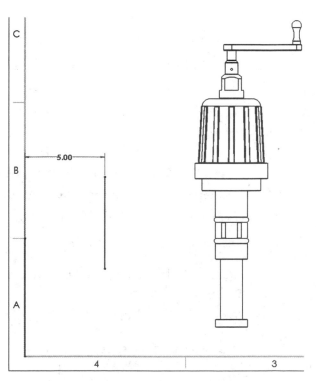

9. Converting an entity:

- Remain in the **Lock View Focus**
  Mode; this way the next entity will
  get converted into a line and belongs
  to the View, <u>not</u> the Sheet.

- Select the vertical edge as noted
  and press the **Convert Entity**
  command from the Sketch tool tab.

- The selected edge is converted to
  a line. When the drawing view is
  moved, the line will move along.

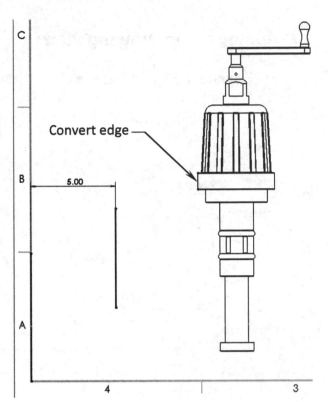

10. Adding a reference dimension:

- The question is: how can we create
  a dimension between a line and an
  edge of a drawing view?

- There are several ways to achieve
  this, and one approach is to lock
  the View-Focus and add the
  reference lines.

- Add a **Driven** dimension as shown.

- Enter the dimension value here:

_____ inches.

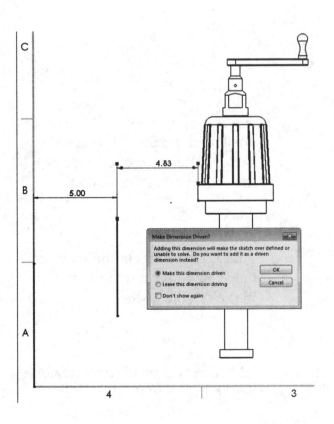

11. Adding a Top drawing view:

- Expand the **View Palette**.

- Drag and drop the **Top** drawing view approximately as shown below.

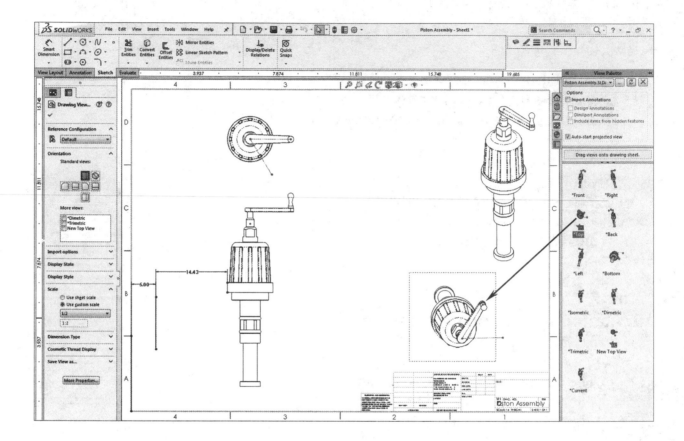

- Switch to the **Sketch tab** and sketch a **line** to the left side of the drawing view as shown.

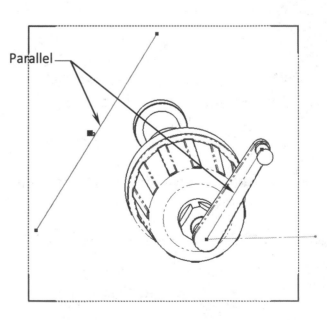

- Add a **Parallel** relation between the sketch line and the centerline in the middle of the crank handle.

- This line will be used to create a Section view in the next step.

- Click the **Section View** command (arrow) from the **View Layout** tab.

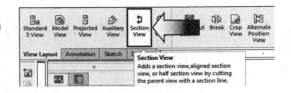

Section View
Adds a section view, aligned section view, or half section view by cutting the parent view with a section line.

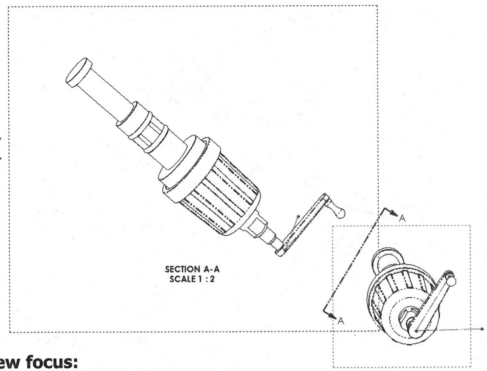

- Place the section view on the upper left side of the Top view.

SECTION A-A
SCALE 1 : 2

12. Locking a view focus:

- Use **Lock-View-Focus** to keep a drawing view active while adding other reference geometry.

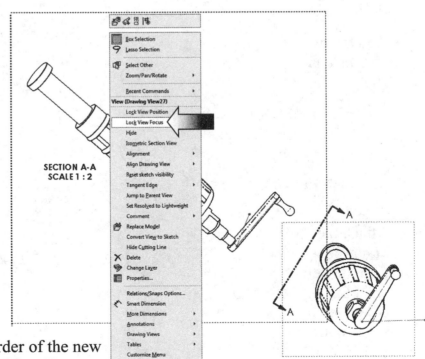

SECTION A-A
SCALE 1 : 2

- The reference geometry can be measured to and from the geometry of the drawing view.

- Right click the border of the new View; select **Lock View Focus**.

- Sketch a new **Line** approximately as shown below.

- Select the horizontal centerline and press **Convert Entities** 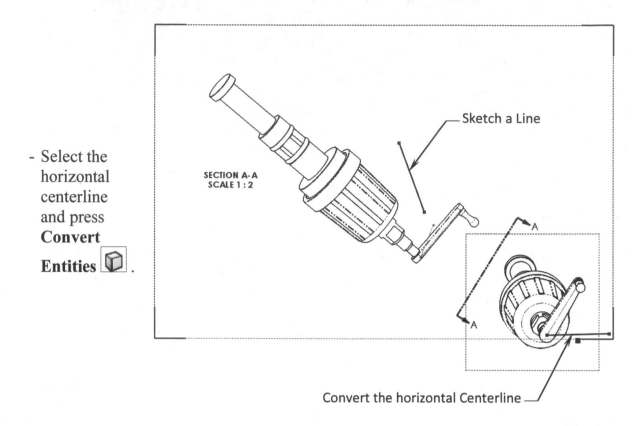.

Sketch a Line

SECTION A-A
SCALE 1 : 2

A

A

Convert the horizontal Centerline

- Add the two dimensions shown.

- Attach the dimensions from the left end of the converted line to the bottom endpoint of the sketched line.

SECTION A-A
SCALE 1 : 2

A

A

8.00

8.00

13. Adding an angular dimension:

- The angular dimension will be used as the answer for this question.

- Add an angular dimension between the sketched line and the left-most edge of the crank handle.

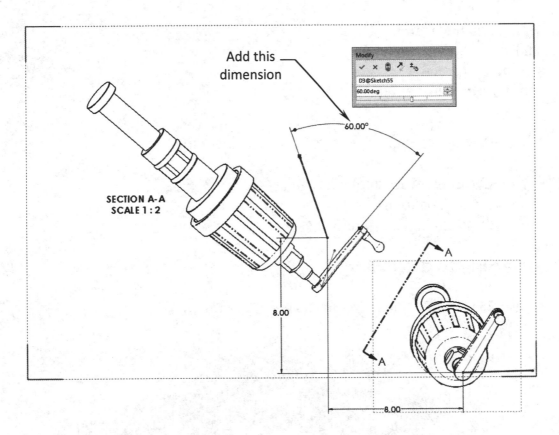

- Enter the dimension value here:

_____ degrees.

Summary:

The key features to the Challenge 2 are:

- Creating the drawing views and finding the right orientations to assist with creating the other drawing views.

- Lock and Un-lock the View Focus so that reference geometry can be added for measuring and locating other references.

CHALLENGE 3

1. Opening an assembly document:

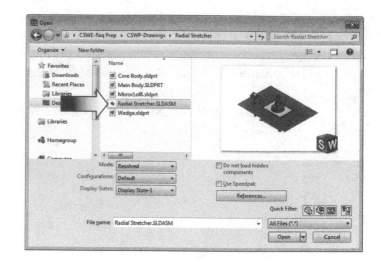

- Select **File / Open**.

- Browse to the Training Folder and open the assembly document named **Radial Stretcher.sldasm**.

- This challenge examines your skills on the following:

  * Creating an assembly drawing.
  * Adding balloons.
  * Customizing a bill of materials.

2. Transferring to a drawing:

- Select **File / Make Drawing from Assembly** (arrow).

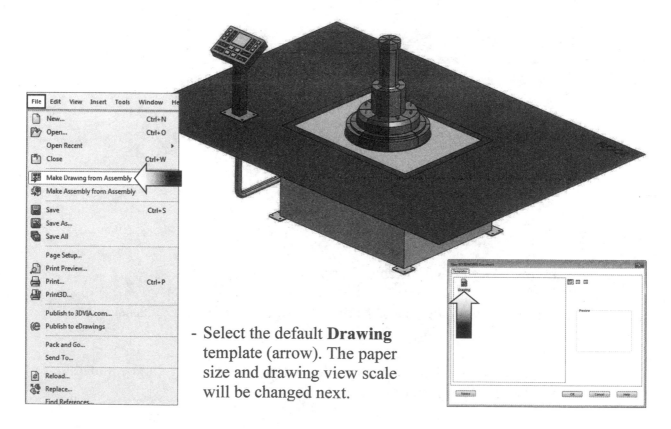

- Select the default **Drawing** template (arrow). The paper size and drawing view scale will be changed next.

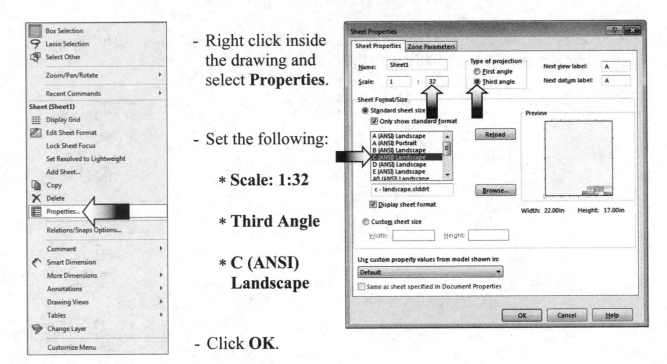

- Right click inside the drawing and select **Properties**.

- Set the following:

    * **Scale: 1:32**

    * **Third Angle**

    * **C (ANSI) Landscape**

- Click **OK**.

3. Adding drawing views from the View Palette:

- Expand the **View Palette** and drag/drop the **Isometric Exploded View** to the drawing.

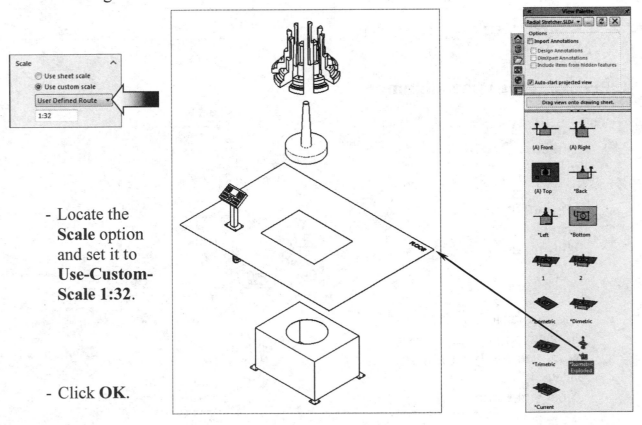

- Locate the **Scale** option and set it to **Use-Custom-Scale 1:32**.

- Click **OK**.

- Next, drag and drop the **Isometric View** also from the View Palette. The drawing view is aligned horizontally with the first view by default.

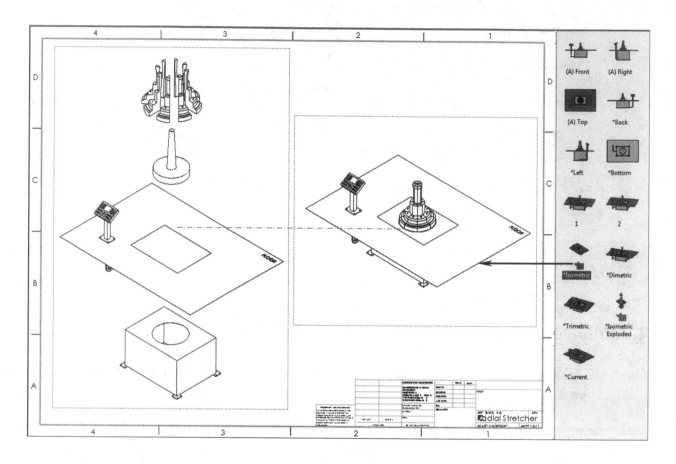

4. Breaking the view alignment:

- Right click the Isometric view's border and select **Alignment / Break-Alignment**.

- The isometric view can now be moved freely. Move it to the upper right side.

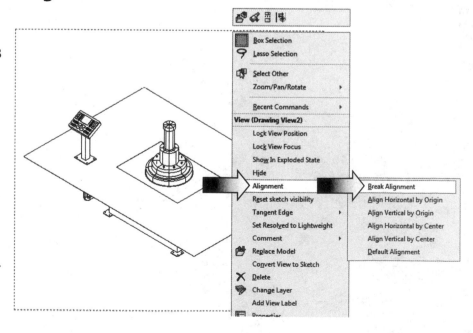

5. Adding balloons:

- Balloons are used to identify the item numbers in the bills of materials.

- Switch to the **Annotation** tab and click the **Auto Balloon** command .

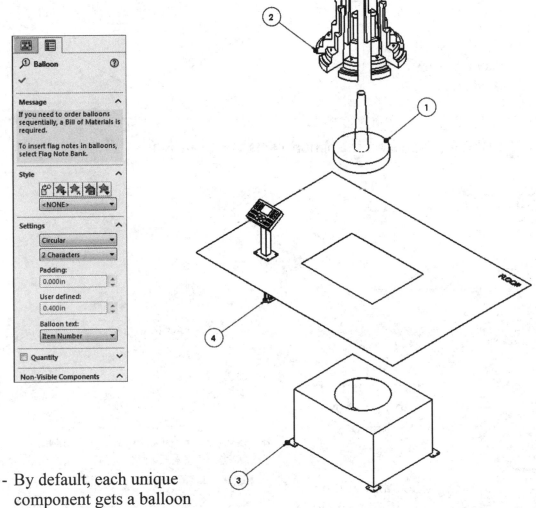

- By default, each unique component gets a balloon assigned to it automatically. Change the balloon settings to **Circular, 2 characters** and click **OK**.

- The item numbers reflect the order of the components listed in the top level assembly. Changes done to the order of the components in the assembly design tree will populate to the balloons and the bill of materials.

6. Adding a bill of materials:

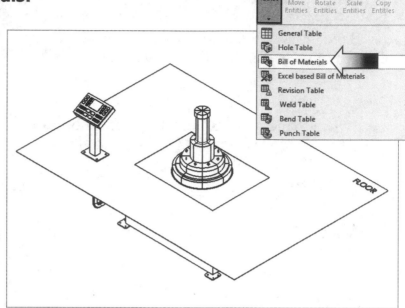

- In an assembly drawing a bill of materials is created to display the item numbers, quantities, part numbers, and custom properties of the assembly.

- From the Annotation tab, select **Tables / Bill of Materials**.

- In the BOM Type, select the option **Parts Only** (arrow).

- Click **OK**.

- Place the Bill of Materials above the title block.

- The table will be modified to include some custom properties in the next couple steps.

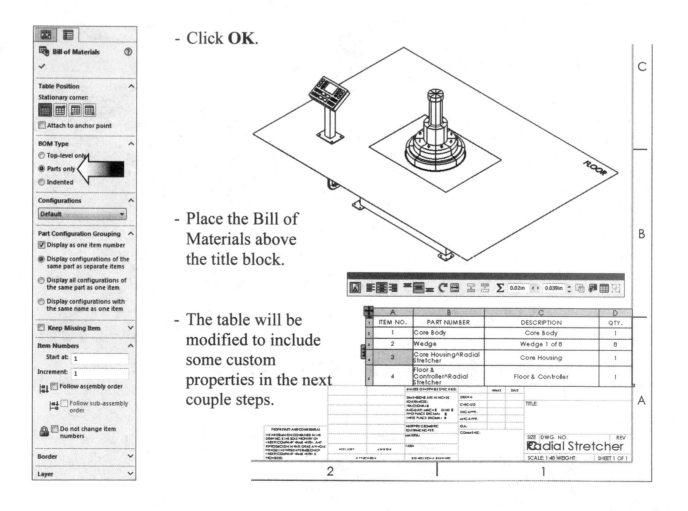

ITEM NO.	PART NUMBER	DESCRIPTION	QTY.
1	Core Body	Core Body	1
2	Wedge	Wedge 1 of 8	8
3	Core Housing^Radial Stretcher	Core Housing	1
4	Floor & Controller^Radial Stretcher	Floor & Controller	1

ITEM NO.	PART NUMBER	DESCRIPTION	QTY.
1	Core Body	Core Body	1
2	Wedge	Wedge 1 of 8	8
3	Core Housing^Radial Stretcher	Core Housing	1
4	Floor & Controller^Radial Stretcher	Floor & Controller	1

- Zoom in on the Bill of materials. We will change the Part Number column to include the actual part numbers that was assigned earlier from the part level.

7. Changing custom properties:

- Double click the column header **B** to access the Custom Property options.

- Change the Column Type to **Custom Property** (arrow).

- For Property Name, select **PartNo** from the list (arrow).

	A	B	C	D
1	ITEM NO.	PartNo	DESCRIPTION	QTY.
2	1	232 178 0313	Core Body	1
3	2	417902661	Wedge 1 of 8	8
4	3	424 514 6229	Core Housing	1
5	4	292 436 5662	Floor & Controller	1

- The part numbers for each component are displayed in column B.

- Adjust the column width by dragging the row divider ⬅||➡ .

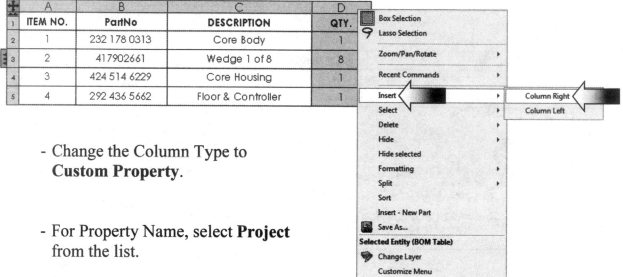

8. Adding a new column:

- Right click the column header **D** and select **Insert / Column Right** (arrow).

- Change the Column Type to
  Custom Property.

- For Property Name, select **Project**
  from the list.

- The project **Radial Stretcher** is displayed
  in the new column.

- The completed Bill of Materials.

ITEM NO.	PartNo	DESCRIPTION	QTY.	Project
1	232 178 0313	Core Body	1	Radial Stretcher
2	417902661	Wedge 1 of 8	8	Radial Stretcher
3	424 514 6229	Core Housing	1	Radial Stretcher
4	292 436 5662	Floor & Controller	1	Radial Stretcher

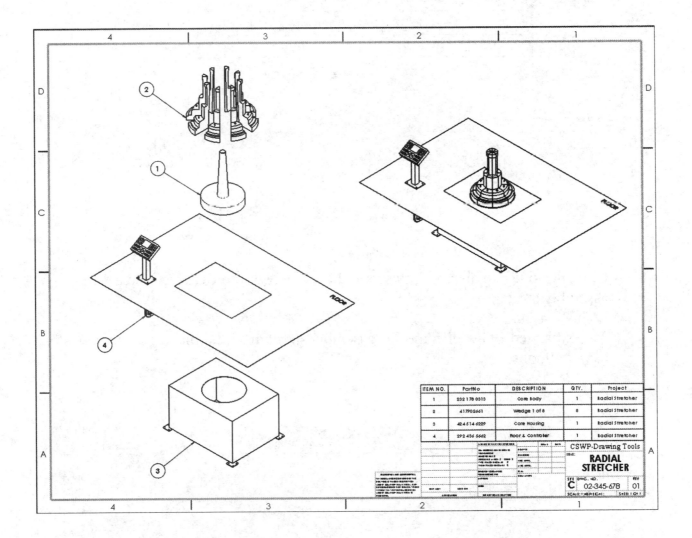

Summary:

The key features to the Challenge 3 are:

- Creating an assembly drawing complete with balloons, bill of materials, and custom properties.

9. Optional:

- You can expand a BOM to view the assembly structure. For models with balloons, the assembly structure column is preceded by a per-component listing of balloons.

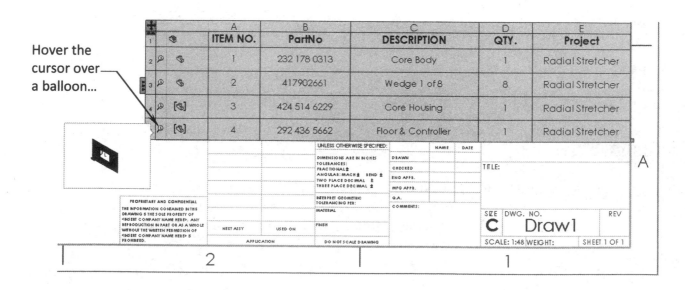

- Click the side expansion arrows ▤ at the left side of the BOM to display the assembly structure.

- The expanded BOM displays the assembly structure and indicates components that have balloons.

CHAPTER 2

Mold Tools

CSWPA - Mold Tools

Certified SOLIDWORKS Professional Advanced Mold Tools

The completion of the Certified SOLIDWORKS Professional Advanced Mold Tools (CSWPA-MT) exam shows that you have successfully demonstrated your ability to use SOLIDWORKS Mold Tools functionality.

Successful completion of this exam will demonstrate knowledge of how to correctly use these SOLIDWORKS tools.

Note: You must use at least SOLIDWORKS 2008 for this exam. Any use of a previous version will result in the inability to open some of the testing files.

Exam Length: 90 minutes

Minimum Passing grade: 80%

Re-test Policy: There is a minimum 30 day waiting period between every attempt of the CSWPA-DT exam. Also, a CSWPA-MT exam credit must be purchased for each exam attempt.

All candidates receive electronic certificates and a personal listing on the CSWP directory when they pass.

Exam features hands-on challenges in many of these areas of SOLIDWORKS Mold Tools functionality such as:

Parting Line Creation, Parting Surface Creation, Draft Analysis, Shut-off Surface Creation, Imported part Repair, Cavity Tool, and Parting Line Split Face.

CSWPA - Mold Tools

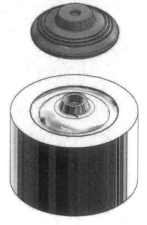

Dimensioning Standards: **ANSI**

Units: **INCHES** – 3 Decimals

Tools Needed:

 Extruded Surface Planar Surface Filled Surface

 Move/Copy Knit Surface Draft Analysis

 Measure Parting lines Parting Surfaces

CHALLENGE 1

1. Opening a part document:

- Select **File / Open**.

- Browse to the Training Folder and open the part document named **Cavity Creation.sldprt**.

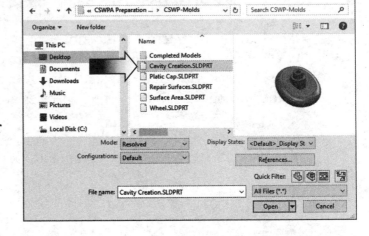

2. Enabling the Mold Tools:

- Right click the **Evaluate** tab and enable the **Mold Tools** (arrow).

- The mold tools span from initial analysis to creating the tooling split. The result of the tooling split is a multibody part containing separate bodies for the molded part, the core, and the cavity. This challenge examines your skills on creating the Cavity half of the mold.

3. Creating the parting Line:

- Click **Parting Lines** ⬡.

- For Direction of Pull, select the **Top** plane from the Feature tree.

- For Draft Angle, enter **1.00 deg**.

- Enable the **Split-Faces** checkbox.

- For Parting Lines, select the **edge** as noted.

- Click **OK**.

Parting Lines
Establishes parting lines to separate core and cavity surfaces.

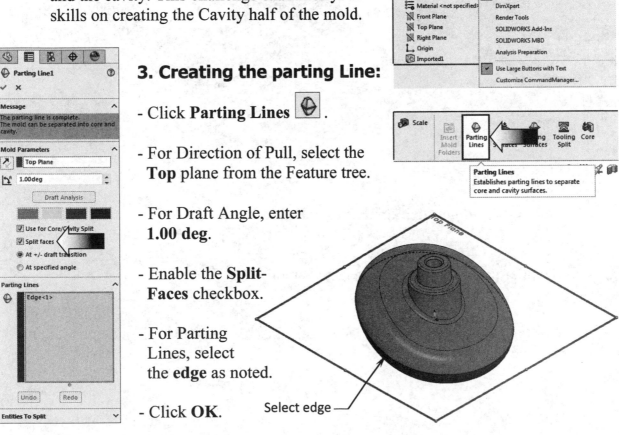

Select edge

4. Creating the Parting Surfaces:

- The Parting surfaces split the mold cavity from the core. The parting lines are created first, the shut off surfaces are created only if there are through hole in the part, and then the parting surfaces are created afterwards.

- Click the **Parting Surfaces** command.

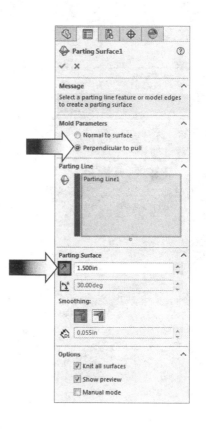
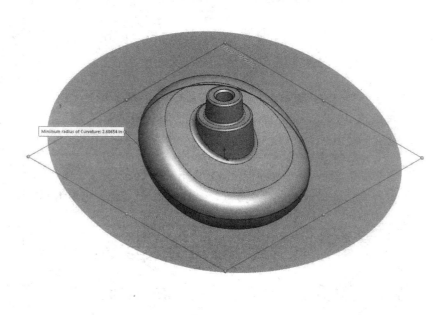

- Select the **Perpendicular to Pull** option (arrow).

- The Parting Line should be selected automatically.

- For Parting Surface, enter **1.500in** and click **Reverse** direction to extend the surface outward.

- Click **OK**.

5. Knitting the Surfaces:

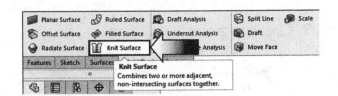

- If the model was created correctly all surfaces will knit automatically.

- When knitting surfaces to within a tolerance and the distance between two surface edges exceeds the tolerance, the resulting knitting gaps are considered open. In the Knit Surface PropertyManager, under Gap Control, you can modify the knitting tolerance to improve the surface knit, based on resulting gaps.

- Click **Knit Surface**.

- Select the **Parting Surface** and the **bottom surface** of the part as noted.

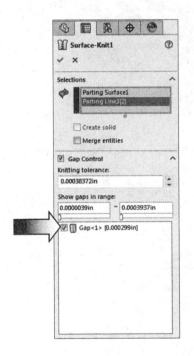

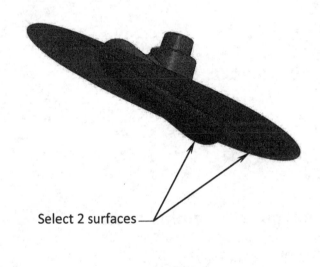

Select 2 surfaces

- Enable the **Gap Control** checkbox and click the keep gap checkbox (arrow).

- Click **OK**.

(Only the Knit Surface is shown here.)

6. Creating a new plane:

- Switch to the **Features** tool tab and select **Reference Geometry / Plane**.

- Select the **Top** plane from the Feature tree.

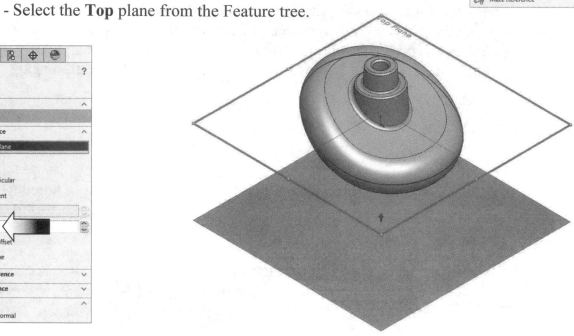

- Enter **2.000in** for Offset Distance and <u>enable</u> the **Flip Offse**t checkbox (arrow).

- Click **OK**.

7. Sketching the mold block:

- Select the <u>new plane</u> and open a **new sketch**.

- Sketch a **Center Rectangle** centered on the origin.

- Add the height and width dimensions to fully define the sketch.

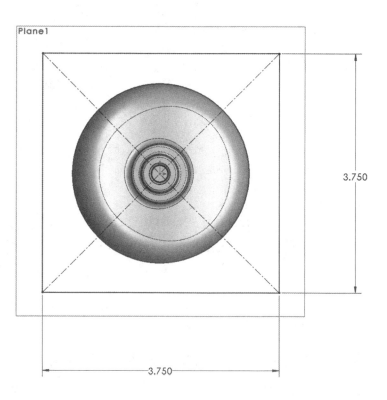

8. Creating the Cavity block:

- Switch to the **Features** tool tab.

- Click **Extruded Boss Base** .

- For Direction 1, select the **Up-To-Surface** option.

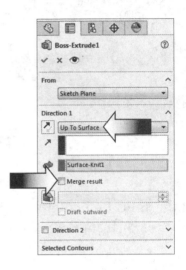

- Select the **Knit Surface** either from the graphics area or from the Feature tree.

- <u>Clear</u> the **Merge Result** checkbox.

- Click **OK**.

9. Separating the block from the part:

- To be able to see the cavity block, we will need to separate it from the part.

- Select **Insert / Features / Move-Copy**.

Select the Knit Surface

- Click the **Translate / Rotate** button (arrow)
  to change to the move options.

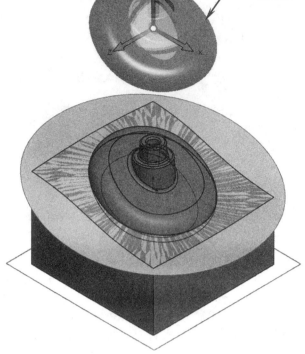

Body to move

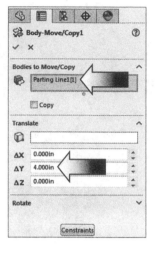

- For Bodies to Move/Copy, select the **solid body** (**Parting Line1[1]**) either from the graphics area or from the Solid Bodies folder.

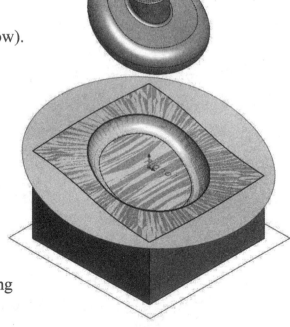

- Enter **4.000in** in the **Delta Y** field (arrow).

- Click **OK**.

- The solid model is moved and the Cavity block is now visible for analyzing or calculating the mass.

10. Hiding the reference surfaces:

- The reference surfaces that were used to create the cavity block can now be hidden.

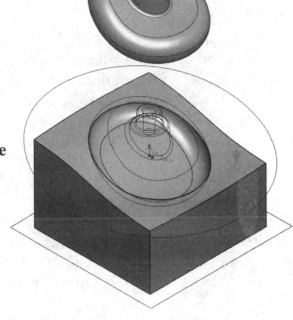

- Right click the **Surface Bodies(2)** folder and select **Hide** (arrow).

11. Measuring the cavity block:

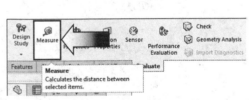

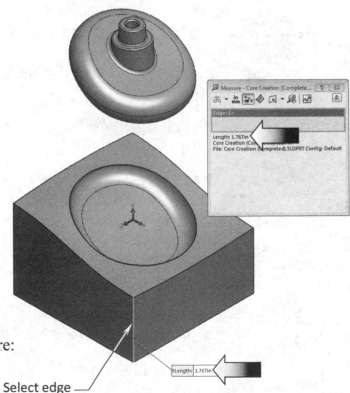

- Switch to the **Evaluate** tab.

- Select the **Measure** tool.

- Select the <u>vertical edge</u> as indicated.

- Enter the <u>length</u> of the edge here:

_____ in.

Select edge

CHALLENGE 2

1. Opening a part document:

- Select **File / Open**.

- Browse to the Training Folder and open the part document named **Plastic Cap.sldprt**.

- This challenge examines your skills on creating only the Core half of the mold.

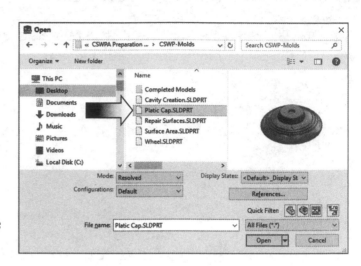

2. Creating the Parting Line:

- Select the **Parting Lines** command 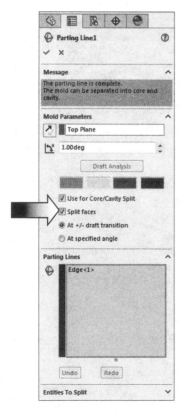 from the **Mold Tools** tab.

- For Direction of Pull, select the **Top** plane from the FeatureManager tree.

- For Draft Angle, enter **1.00deg**.

- Enable the **Split Faces** checkbox.

- For Parting lines, select the **edge** as indicated.

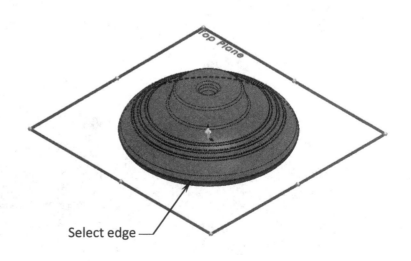

Select edge

- Click **OK**.

3. Creating the Parting Surfaces:

- A shut-off surface is created before the parting surfaces. The command closes up a through holes by creating surface patches along a parting line or edges that form a continuous loop. Since this model does not have any through holes, we will proceed to the next step and create the parting surfaces.

- Click the **Parting Surfaces** command from the Mold Tools tab.

- Select the **Perpendicular to Pull** option (arrow).

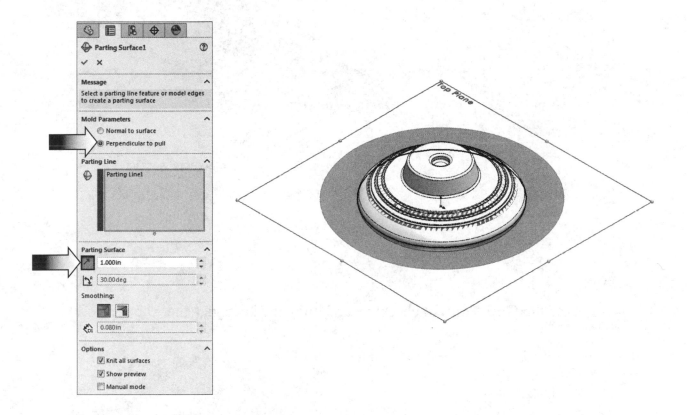

- For Parting Surface, enter **1.000in** for distance and click **Reverse** direction to extend the surface outward.

- Keep other parameters at their default values.

- Click **OK**.

4. Knitting the Surfaces:

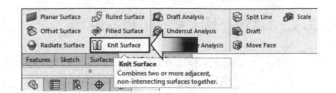

- Click **Knit Surface** 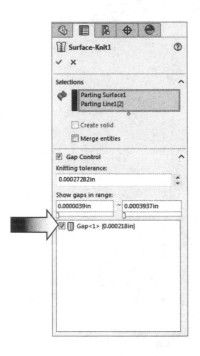.

- Select the **Parting Surface** and the **bottom surface** of the part as noted.

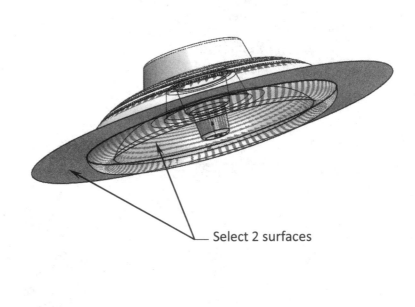

Select 2 surfaces

- Enable the **Gap Control** checkbox and click the keep gap checkbox (arrow).

- Click **OK**.

- The solid model and other reference surfaces are hidden to show only the Knit Surface.

5. Creating a new plane:

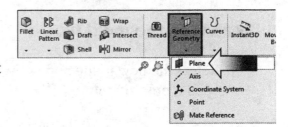

- Switch to the **Features** tool tab and select
  Reference Geometry / Plane.

- For the first reference, select the **Top** plane from the FeatureManager tree.

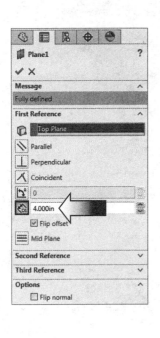

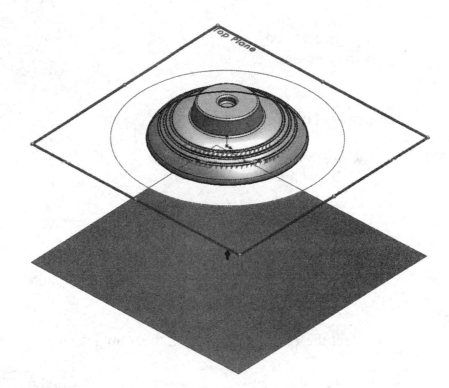

- For Offset Distance, enter **4.000in**.

- Enable the **Flip Offset** checkbox to place the new plane below the Top plane.

- Click **OK**.

6. Sketching the mold block:

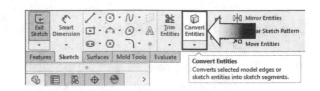

- Select the <u>new plane</u> and open a **new sketch**.

- Select the outer circular edge of the Parting Surfaces and press **Convert Entities**.

- The selected edge is converted into a circle.

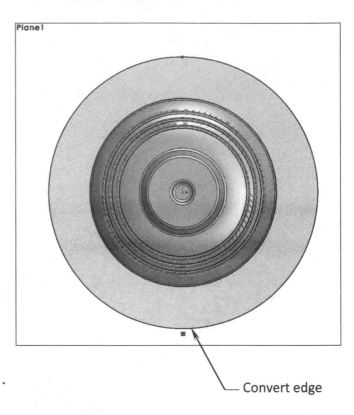

7. Extruding the Core block:

- Switch to the **Features** tool tab.

- Click **Extruded Boss Base** .

Convert edge

- For Direction 1, select the **Up-To-Surface** option.

- Select the **Knit Surface** from the graphics area.

Select the Knit Surface

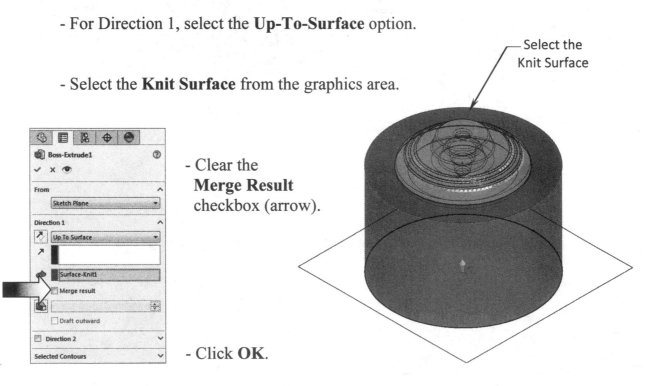

- Clear the **Merge Result** checkbox (arrow).

- Click **OK**.

8. Separating the block from the part:

- The Core needs to be separated from the part.

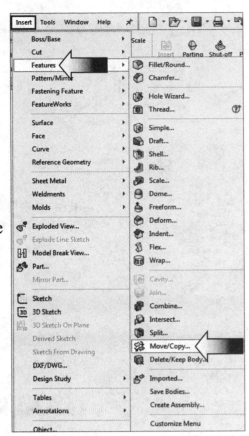

- Select **Insert / Features / Move-Copy**.

- Click the **Translate / Rotate** button to change to the move options.

- For Bodies to Move/Copy, select the **solid body (Parting Line1[1])** either from the graphics area or from the Solid Bodies folder.

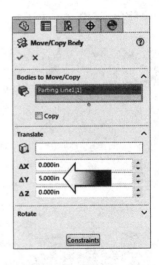

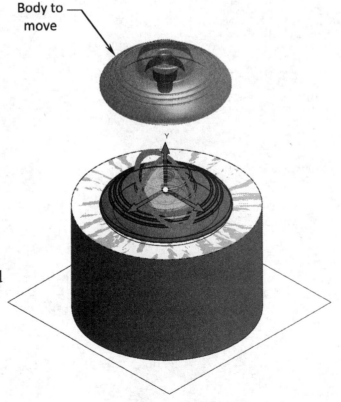

Body to move

- Enter **5.000in** in the **Delta Y** field (arrow).

- Click **OK**.

9. Hiding the reference surfaces:

- The reference surfaces that were used to create the cavity block can now be hidden.

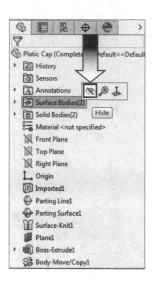

- Right click the **Surface Bodies(2)** folder and select **Hide** (arrow).

10. Assigning material:

- Select the **Chrome Stainless Steel** material for the Core.

- Click **OK**.

11. Calculating the mass:

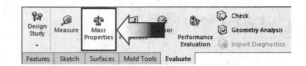

- Switch to the **Evaluate** tab.

- Click **Mass Properties** and select the **Core** block either from the graphics area or from the Solid Bodies folder to calculate the mass.

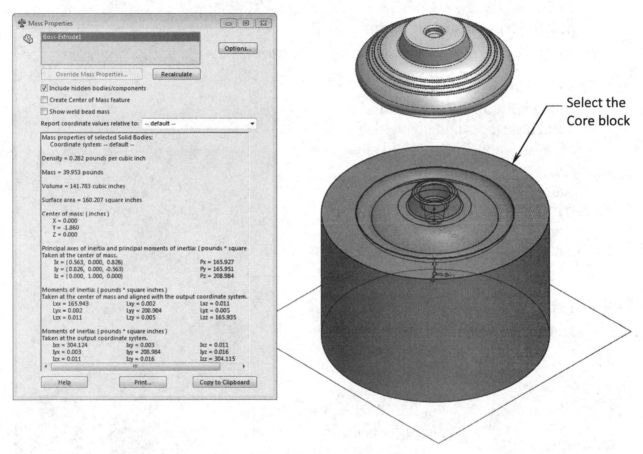

Select the Core block

- Enter the final mass of the Core block here:

_____ pounds.

12. Saving your work:

- Save your model as **Plastic Cap (Completed).sldprt**.

- Close all documents.

CHALLENGE 3

1. Opening a part document:

- Select **File / Open**.

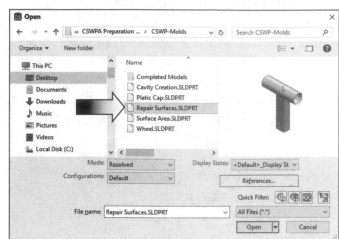

- Browse to the Training Folder
and open the part document
named **Repair Surfaces.sldprt**.

- This challenge reinforces the
following skills: Repairing the
broken surfaces and converting
a surface model into a solid model.

2. Creating a new plane:

- The overall length of the round tube is **5.00in**.

- Click the **Plane** command from the
Reference Geometry drop-down.

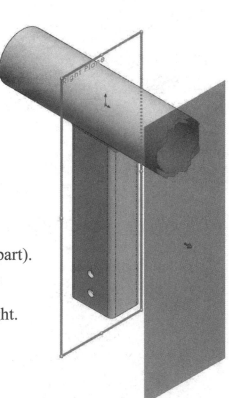

- For the 1st reference, select
the Right plane from the tree.

- For Offset Distance, enter
2.500in (1/2 the width of the part).

- Place the new plane on the right.

- Click **OK**.

3. Converting the geometry:

- Select the <u>new plane</u> and open a **new sketch**.

- Select the **2 circular edges** on the left side of the round tube and press **Convert Entities**.

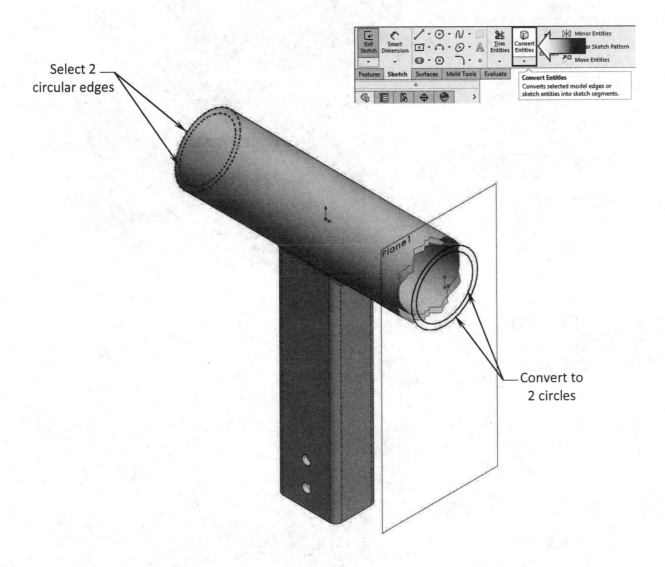

Select 2 circular edges

Convert to 2 circles

- The 2 selected edges are converted into 2 circles.

- The circles will be used to repair the broken surfaces in the round tube.

4. Extruding a surface:

- Switch to the **Surface** tab.

- Click **Extruded Surface** (arrow).

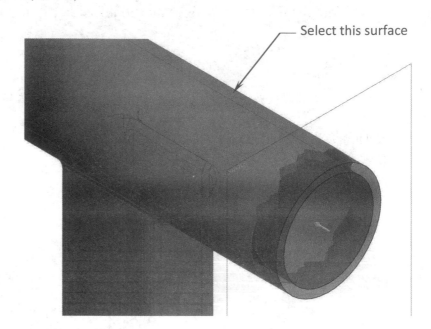

Select this surface

- Select the **Up-To-Surface** option.

- Click the round tube in the graphics area.

- Click **OK**.

- The broken surfaces are filled with a new surface.

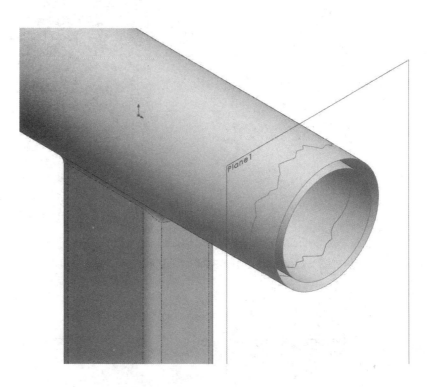

5. Creating a planar surface:

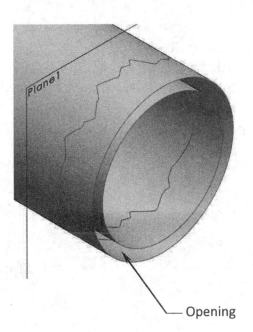

- The extruded surface has an opening on the right end. We need to patch it up with a planar surface.

- Switch to the **Surfaces** tab.

- Click the **Planar Surface** command.

Opening

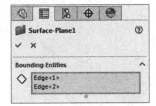

- Select the **2 edges** on the right end of the round tube.

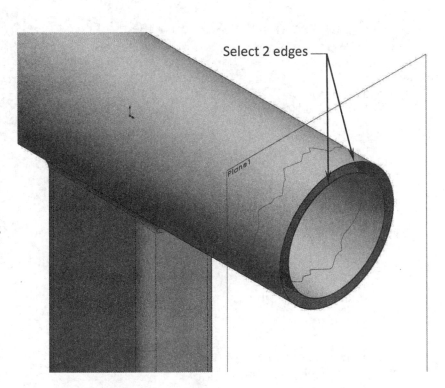

Select 2 edges

- A preview graphics appears displaying a planar surface is being created.

- Click **OK**.

6. Knitting the surfaces:

- Click the **Knit Surface** command .

- Select **all surfaces** in the Surface Bodies folder: Surface-Extrude, Surface-Imported, Surface-Extrude1, Surface-Extrude2, and Surface-Plane.

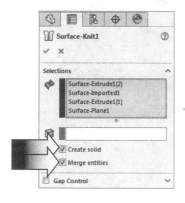

- Enable the **Create Solid** checkbox (arrow).

- Enable the **Merge Entities** checkbox (arrow).

- Click **OK**.

7. Creating a section view:

- Create a section view to verify the thickness of this solid model.

- Click the **Section View** command.

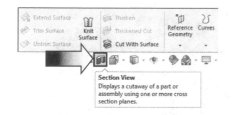

- Use the default **Front** plane as the cutting plane.

- The Blue color represents the thickness of the model.

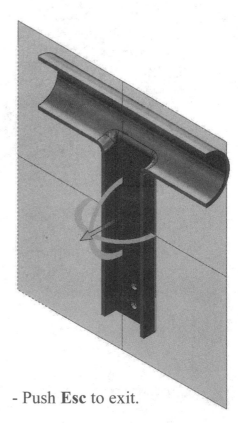

- Push **Esc** to exit.

8. Applying material:

- Right click the **Material** option and select:
  1060 Alloy (arrow).

9. Calculating the mass:

- Switch to the **Evaluate** tab.

- Click **Mass Properties**  .

- Enter the final mass of the model here:

_____ pounds.

- <u>**Save**</u> your model as **Repair Surfaces (Completed).sldprt** and close all documents.

CHALLENGE 4

1. Opening a part document:

- Select **File / Open**.

- Browse to the Training Folder and open the part document named **Surfaces Area.sldprt**.

- This challenge examines your skills on filling the openings and measuring the total surface area of the patches.

2. Creating the Filled Surfaces:

- One of the steps in preparing for a mold design is to fill all openings either automatically with the Shut-Off Surfaces command, or to manually patch them one by one. For this challenge, we are going to fill each opening manually.

- Click the **Filled Surface** command. Select the **6 bottom edges** of one of the slots.

- For Curvature Control, select **Tangent** and enable the **Apply to All Edges** box.

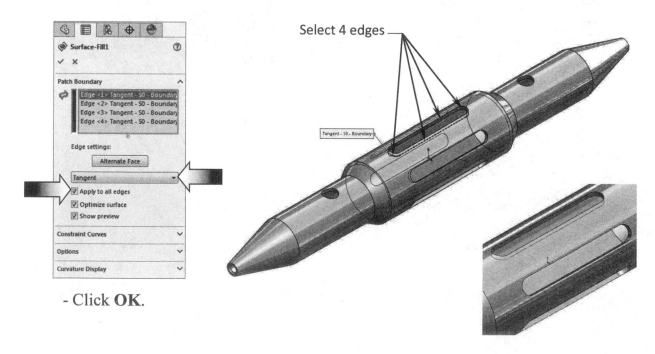

- Click **OK**.

3. Filling other slots:

- Repeat the step 2 and fill the other slots using the same settings as the 1st one.

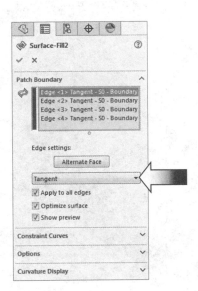

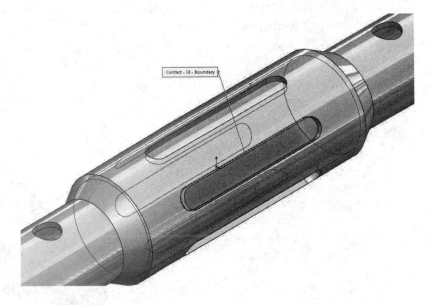

4. Filling the holes:

- The Tangent control may not work very well with the holes, use **Contact** instead.

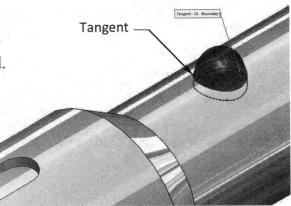

- Fill the 4 holes one at a time, use the same **Contact** control (arrow).

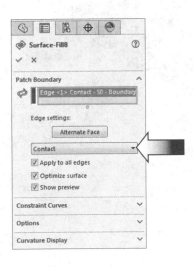

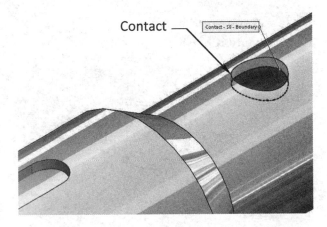

- Click **OK**.

5. Filling the ends with Planar Surfaces:

- Use the Filled Surface command to patch non-planar openings, but use the Planar Surface command to patch planar openings. The 2 ends of this model can be patched with Planar Surfaces.

- Click the **Planar Surface** command 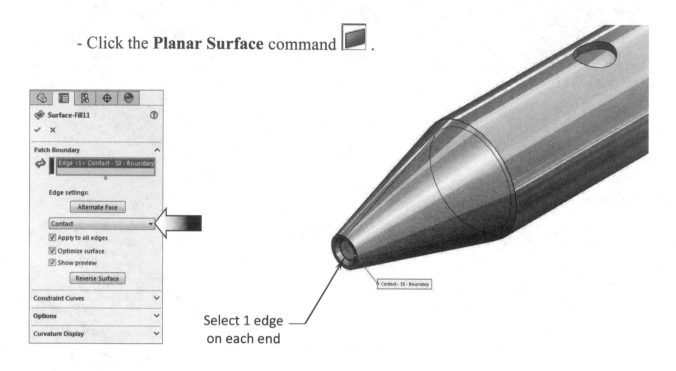 .

Select 1 edge on each end

- Select the **inside edge** on each end of the model.

- For Curvature Control, use the **Contact** option.

- Click **OK**.

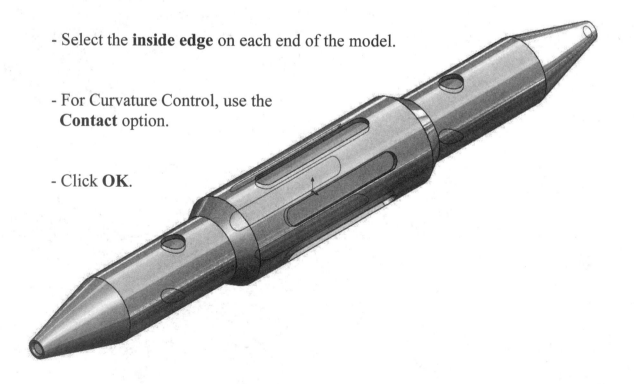

6. Measuring the total surface area of the patches:

- Switch to the **Evaluate** tab.

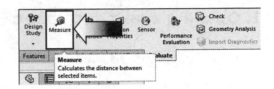

- Click the **Measure** command 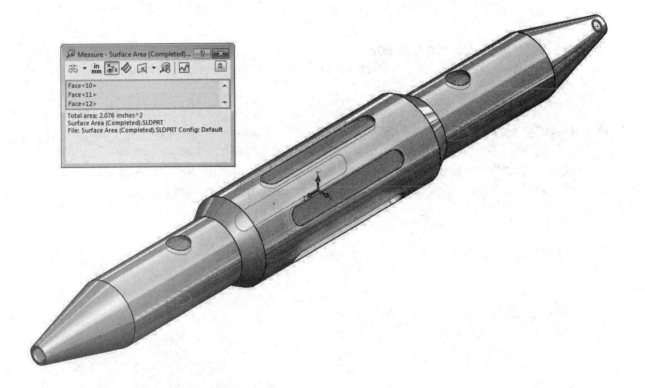.

- Select all **12 surfaces** in the graphics area.

- Enter the total surface area here:

_____ inches^2

7. Saving your work:

- <u>**Save**</u> your work as **Surface Area (Completed).sldprt**.

- Close all documents.

CHALLENGE 5

1. Opening a part document:

- Select **File / Open**.

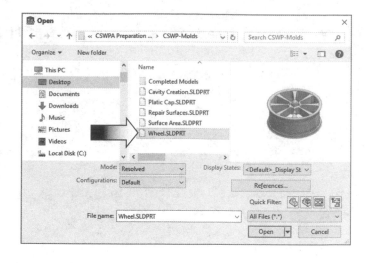

- Browse to the Training Folder
  and open the part document
  named **Wheel.sldprt**.

- This challenge examines your
  skills on analyzing the positive and negative draft angles in the model.

2. Analyzing the Draft Angles:

- Use Draft Analysis to verify draft angles, examine angle changes within a face,
  as well as locate parting lines, injection, and ejection surfaces in parts.

- The **Green** color represents faces with **Positive** draft.

- The **Yellow** color represents the faces
  that have no draft or **Required** draft.

- The **Red** color represents the
  Faces with **Negative** draft.

- The **Blue** color represents the
  Straddle faces that contain
  both positive and negative draft.

- Select the **Draft Analysis**
  command from the **Surfaces** tab.

- For Direction of Pull, select the **Top** plane from the FeatureManager tree.

- For Draft Angle, enter **3.00deg**.

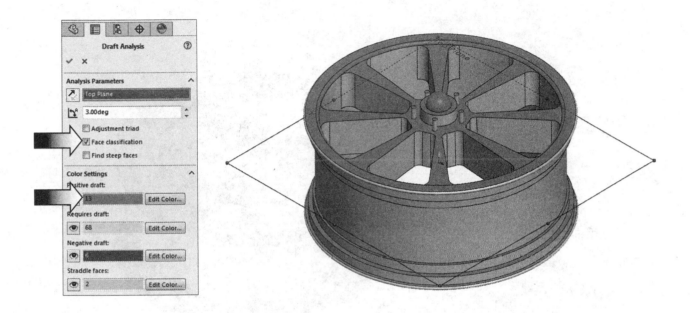

- Click the **Face Classification** checkbox (arrow).

- Enter the number of **Positive Draft** (green faces) here:

_____ Positive Draft Faces.

- Remain in the Draft Analysis mode, we will change the draft angle and analyze the model again.

3. Changing the Draft Angle:

- Change the Draft Angle to **1.00deg** (arrow).

- Keep the **Face Classification** checkbox enables.

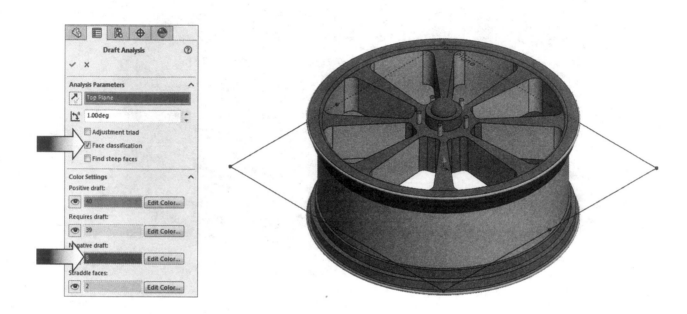

- Enter the number of **Negative Draft** faces (red color) here:

_____ Negative Draft Faces.

4. Saving your work:

- **Save** your work as **Wheel (Completed).sldprt**.

- Close all documents.

CHAPTER 3

Weldments

PROFESSIONAL
Weldments
SOLIDWORKS

CSWPA – Weldments

The completion of the Certified SOLIDWORKS Professional
Advanced Weldments (CSWPA-WD) exam proves that you have successfully
demonstrated your ability to use the SOLIDWORKS tools for Weldments.

Employers can be confident that you understand the SOLIDWORKS tools that will
aid in the design of Weldment components.
Recommended Training Courses: Weldments

Note: *You must use at least SOLIDWORKS 2010 for this exam. Any use of a
previous version will result in the inability to open some of the testing files.*

Exam Length: 2 hours
Minimum Passing grade: 75%

Re-test Policy: There is a minimum 30 day waiting period between every attempt
of the CSWPA-WD exam. Also, a CSWPA-WD exam credit must be purchased
for each exam attempt.

All candidates receive electronic certificates and a personal listing on the CSWP
directory when they pass.

Exam features hands-on challenges in many of these areas of SOLIDWORKS
Weldment functionality:

Weldment Profile creation, Placing Weldment Profile in the Weldment Library,
Basic and Advanced Weldment Part Creation and modification, Adding End
Caps and Gussets, Using Trim & Extend command, 3D Sketch creation, Cut
List Management, and Cut List Creation in Weldment Drawings.

CSWPA – Weldments

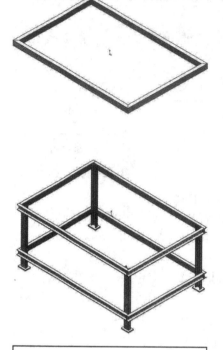

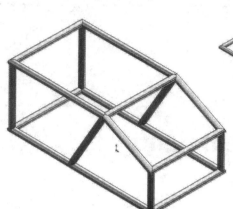

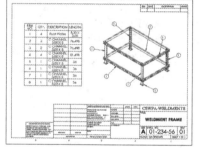

View Orientation Hot Keys:

Ctrl + 1 = Front View
Ctrl + 2 = Back View
Ctrl + 3 = Left View
Ctrl + 4 = Right View
Ctrl + 5 = Top View
Ctrl + 6 = Bottom View
Ctrl + 7 = Isometric View
Ctrl + 8 = Normal To Selection

Dimensioning Standards: **ANSI**

Units: **INCHES** – 3 Decimals

Tools Needed:

Weldment

Rectangle

Circle

End Cap

Structural Member

Trim/Extend

Cut List

Drawing

Mass Properties

CHALLENGE 1

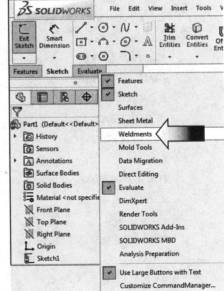

- This challenge focuses on creating the weldment profiles and saving them in the Weldment Profile Library.

1. Starting a new part document:

- Click **File / New**.

- Select the default **Part Template** and set the Units to **Inches, 3 decimal places**.

- Weldments functionality enables you to design a weldment structure as a single multibody part. You use 2D and 3D sketches to define the basic framework. Then you create structural members containing groups of sketch segments. You can also add items such as gussets and end caps using tools on the Weldments toolbar.

2. Enabling the Weldments tab:

- Right click on the **Evaluate** tab and select the **Weldments** tab (arrow).

- When you create the first structural member in a part, a weldment feature is created (arrow) and added to the FeatureManager design tree.

SOLIDWORKS also creates two default configurations in the ConfigurationManager: a parent configuration Default (As Machined) and a derived Configuration Default (As Welded).

- The next step is to create the first weldment profile and save it in the Weldment profiles Library.

3. Sketching the 1st weldment profile:

- Select the **Front** plane and open a new sketch.

- Sketch the **2 rectangles** and add the dimensions shown to fully define the sketch

- Add the **Symmetric** relations where needed to reduce some of the dimensions.

- Add the **Sketch Point** as noted to help locating the profile later on.

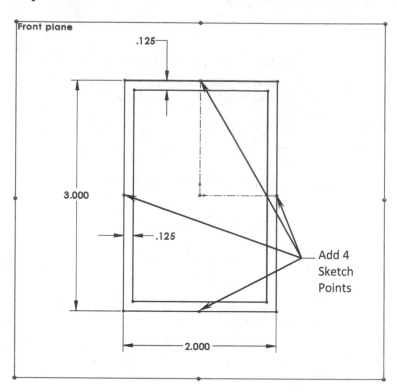

4. Saving the profile:

- **Exit** the Sketch and save the profile under the following directories:

C: Program Files/Solid-Works Corp/ SolidWorks/ Lang/English/ Weldment Profiles/Ansi Inch/ Rectangular Tube.

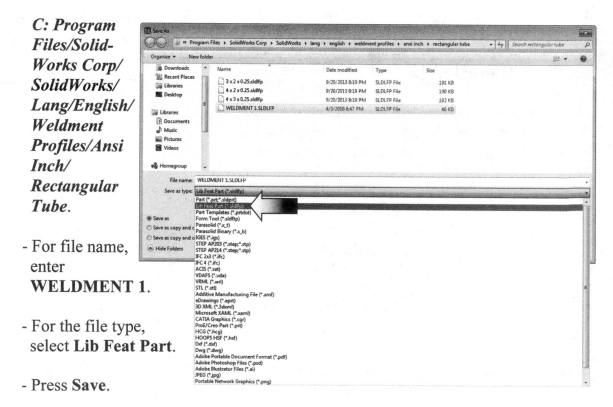

- For file name, enter **WELDMENT 1**.

- For the file type, select **Lib Feat Part**.

- Press **Save**.

5. Creating the 2nd weldment profile:

- Click **File / New / Part**.

- Select the default **Part Template** and set the Units to **Inches, 3 decimal places**.

6. Sketching the profile:

- Open a new sketch on the **Front** plane.

- Sketch **2 circles** and add the diameter dimensions to fully define the sketch.

- Add **4 Sketch Points** as noted to help locate the profile later on.

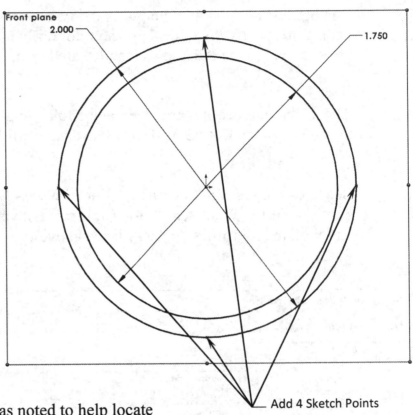

Add 4 Sketch Points

7. Saving the profile:

- **Exit** the sketch and save the profile in the same directory as the last one.

- For the file name, enter **WELDMENT 2**.

- For the file type, select **Lib Feat Part**.

- Press **Save** and close all documents.

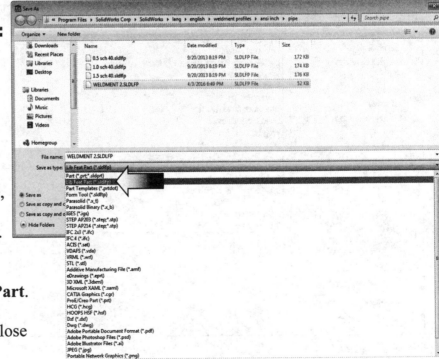

8. Showing the Weldment Profiles Folder:

- The option "Show Folder For" lists file types for which you can define a search path. You specify the folders to search for different types of document. Folders are searched in the order in which they are listed.

- The Weldment Profiles are saved in the following directories: C:\Program Files\ SOLIDWORKS Corp\SOLIDWORKS\lang\english\weldment profiles.

- Select **Options / File Locations / Show Folders For / Weldment Profiles** (arrows) (if the Weldment Profiles folder is not visible, click **Add** and browse to the Weldment Profiles directory listed above).

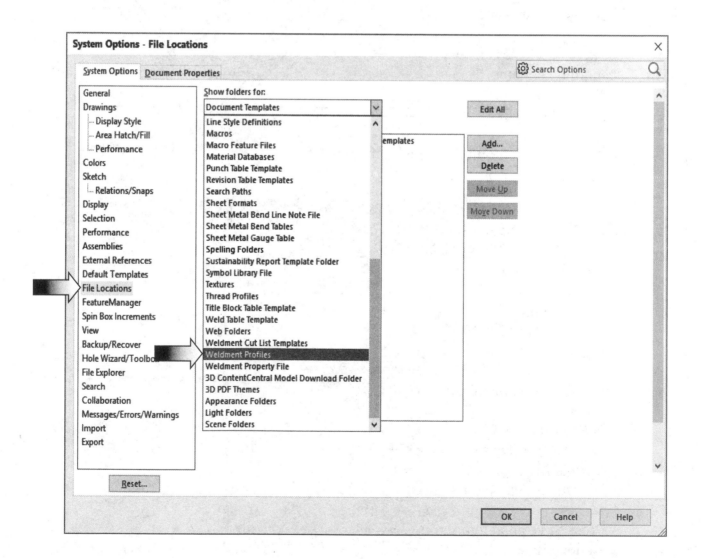

9. Creating a new weldment part:

- Click **File / New / Part**.

- Select the default **Part Template** and set the Units to **Inches, 3 decimal places**.

- Select the **Top** plane and open a new sketch.

- Sketch a **Center Rectangle** as shown and add the height and width dimensions.

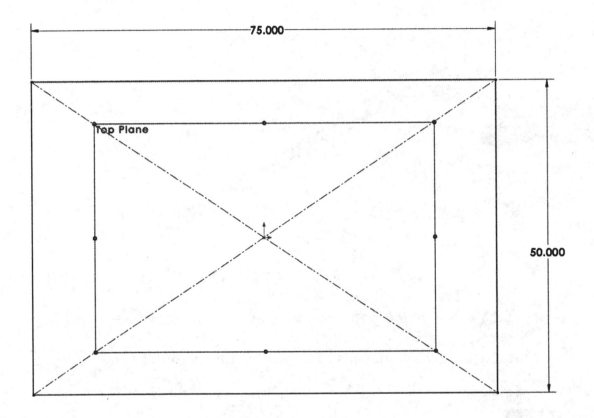

- **Exit** the sketch.

- Click the **Weldment** command (arrow) to add the Weldment and the Cut List features to the FeatureManager tree.

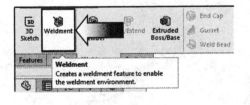

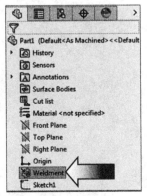

- Two default configurations are added in the Configuration Manager: a parent configuration Default [As Machined] and a derived configuration Default [As Welded].

10. Adding Structural Members:

- The weldment profiles that were created earlier
  will be used to create the weldment structural members.

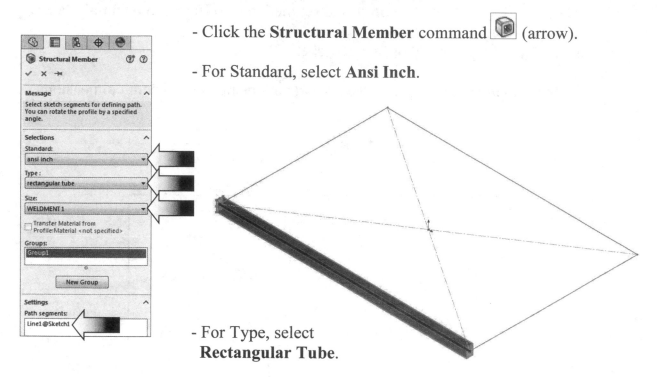

- Click the **Structural Member** command (arrow).

- For Standard, select **Ansi Inch**.

- For Type, select
  Rectangular Tube.

- For Size, select **WELDMENT1**.

- Select the **4 lines** in the graphics area.

- For Apply Corner Treatment, use the default **End Miter** option.

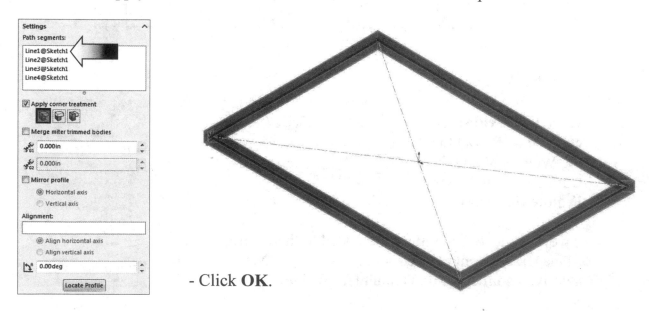

- Click **OK**.

11. Applying material:

- You can apply a material to a part, one or more bodies of a multibody part. In the FeatureManager design tree, in the Solid Bodies folder, right-click a body and click Material. To affect several bodies, select them before right-clicking.

- Right click the Material option on the Feature Manager tree (arrow) and select **Plain Carbon Steel**.

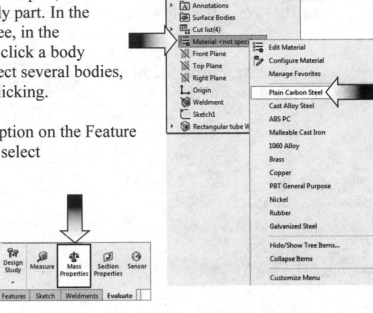

12. Calculating the mass:

- Click **Mass Properties** on the Evaluate tab.

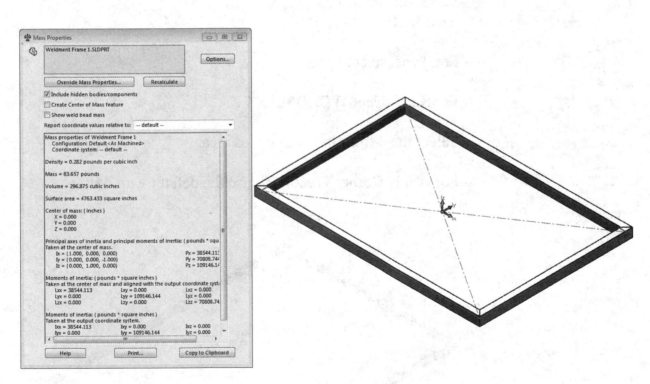

- Enter the mass here _____ Pounds.

- <u>Save</u> the part file as **Weldment Frame 1** but keep the part file open.

13. Changing the Structural Members:

- **Delete** the Structural Member feature from the FeatureManager tree but keep the Sketch1. We will apply the 2nd weldment profile to it.

- Click the **Structural Member** 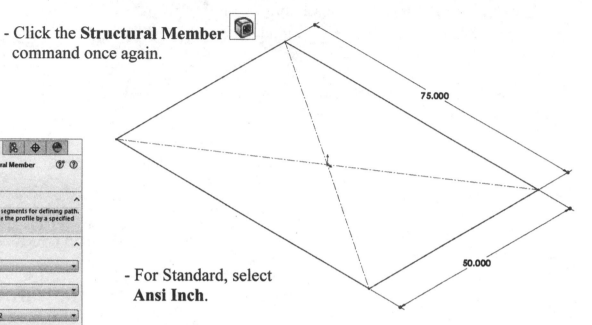 command once again.

- For Standard, select **Ansi Inch**.

- For Type, select **Pipe**.

- For Size, select **WELDMENT 2**.

- Select the **4 lines** in the graphics area.

- For Apply Corner Treatment, use the default **End Miter.**

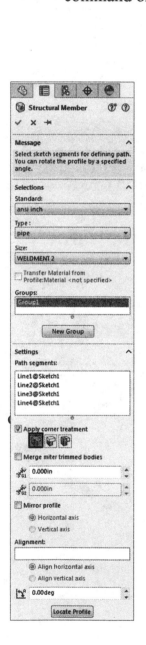

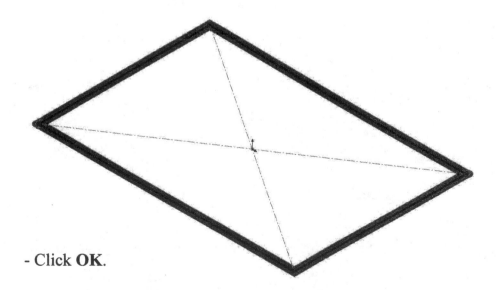

- Click **OK**.

14. Calculating the mass:

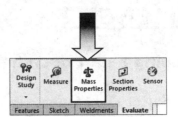

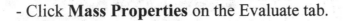

- Use the same material as the last weldment part: **Plain Carbon Steel**.

- Click **Mass Properties** on the Evaluate tab.

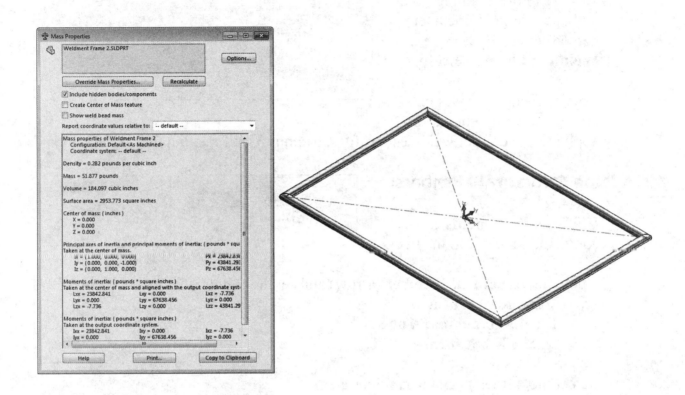

- Enter the mass here _____ Pounds.

- **Save** the part file as **Weldment Frame 2**.

- Close all documents.

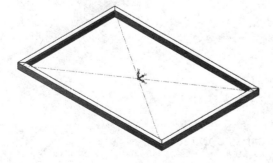

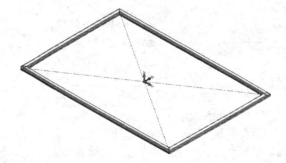

CHALLENGE 2

1. Opening a part document:

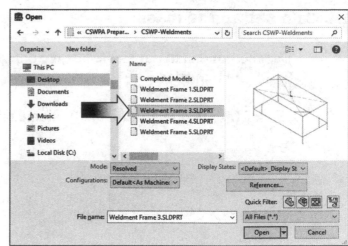

- Select **File / Open**.

- Browse to the Training Folder and open the part document named **Weldment Frame 3.sldprt**.

- This challenge focuses on creating a weldment part by applying the correct weldment profiles, trimming, and treating the corners.

2. Adding Structural Members:

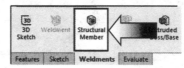

- Click the **Weldments** command to add it to the tree.

- Click the **Structural Member** (arrow) and set the following:
    Standard: **Ansi Inch**.
    Type: **Rectangular Tube**.
    Size: **3 X 2 X 0.25**.

- Select the **4 lines** from the graphics area.

- Click **OK**.

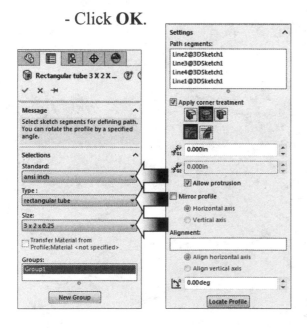

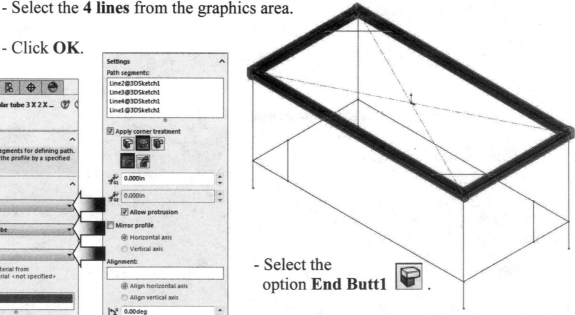

- Select the option **End Butt1** 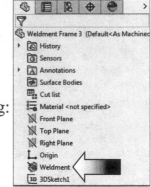.

- Group 1 is created.

- A group is a collection of related segments in a structural member. You configure a group to affect all its segments without affecting other segments or groups in the structural member.

- There are 2 types of groups:

  - **Contiguous Group**: A continuous contour of segments joined end-to-end. You can control how the segments join to each other. The end point of the group can optionally connect to its beginning point.

  - **Parallel Group**: A discontinuous collection of parallel segments. Segments in the group cannot touch each other.

3. Adding Structural Members to Group 2:

- Click the **Structural Member** command 🔲 again.

- Use the same selections as the Group 1 and select the **4 lines** below the Group 1.

- Keep the Corner Treatment at its default settings, including the **End Butt 1** 🔲.

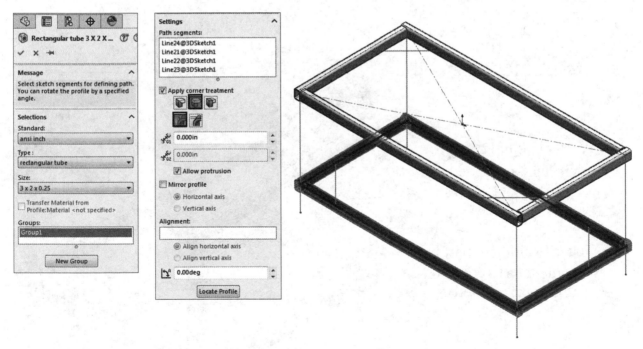

- Click **OK**.

- Group 2 is created.

4. Creating the Parallel Group:

- Click the **Structural Member** command 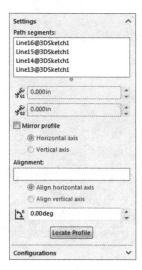.

- Use the <u>same selections</u> as the last 2 groups.

- Select the **4 vertical lines** as shown.

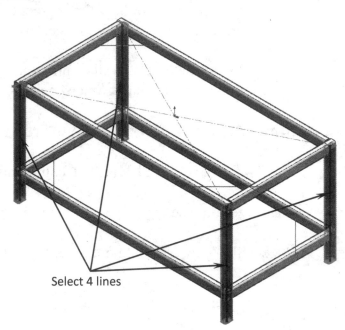

Select 4 lines

- Click **OK**.

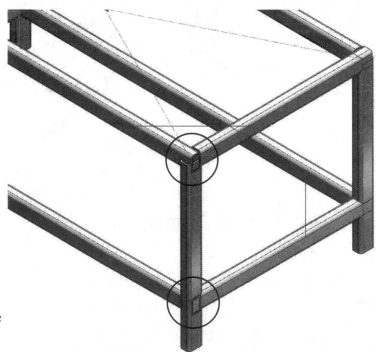

- The Parallel Group, Group 3 is created.

- The weldment part has some interference at this point (circled).

- The tubes will be trimmed to the correct lengths in the next couple steps.

5. Trimming the Parallel Group:

- Click the **Trim/Extend** command 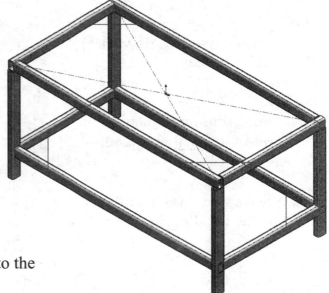 .

- For Corner Type, use the default **End Trim** .

- For Bodies to be Trimmed, select the **4 vertical tubes**.

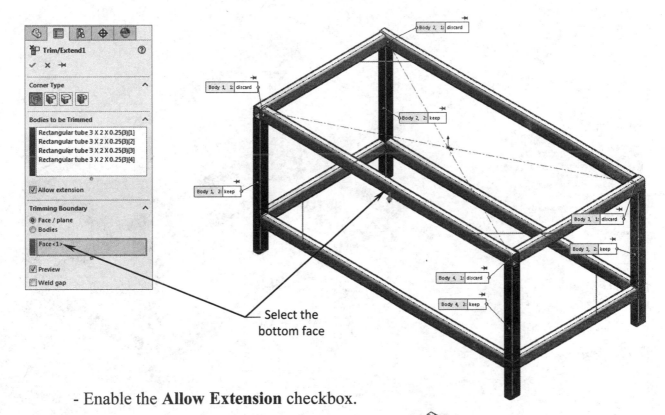

Select the
bottom face

- Enable the **Allow Extension** checkbox.

- For Trimming Boundary, select
the **Face / Plane** option.

- Select the **bottom face** of the
long-horizontal tube as noted.

- Click **OK**.

- The 4 vertical tubes are trimmed to the
correct length.

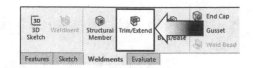

6. Trimming the Group 2:

- Click the **Trim/Extend** command 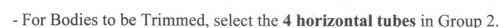.

- For Corner Type, use the default **End Trim** .

- For Bodies to be Trimmed, select the **4 horizontal tubes** in Group 2.

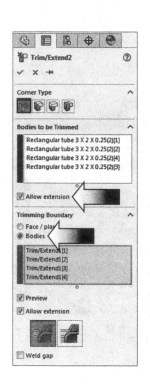

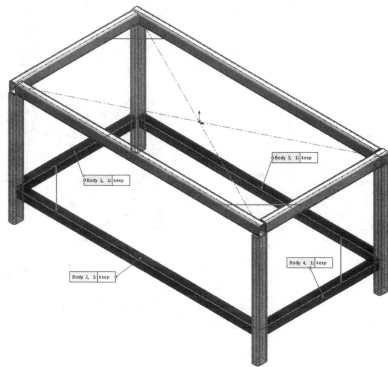

- Enable the **Allow Extension** checkbox (arrow).

- For Trimming Boundary, select the **Bodies** option.

- Select the **4 vertical tubes** in the Parallel Group.

- Click **OK**.

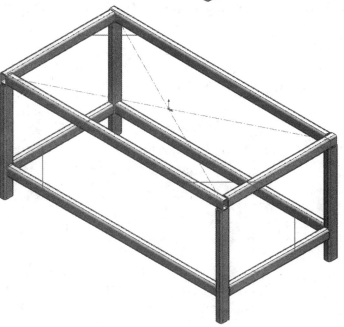

7. Applying material:

- Right click the Material option on the Feature Manager tree and select **Plain Carbon Steel**.

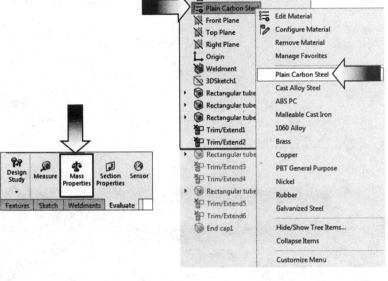

8. Calculating the mass:

- Click **Mass Properties** on the Evaluate tab.

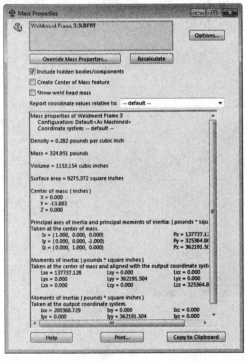

- Enter the mass here: _____ Pounds.

9. Creating the Upper Braces:

- Click the **Structural Member** command .

- Use the <u>same selections</u> as the other 3 groups.

- Select the **2 angled lines** on the <u>top</u> of the main sketch.

- Click **OK**.

10. Trimming the first Upper Brace:

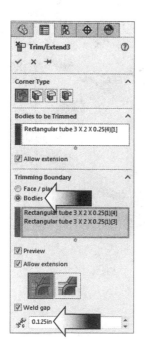

- Trim the Upper Brace using the **End Trim** and **Bodies** options.

- Enable the **Weld Gap** checkbox (arrow) and enter **.125in** for gap.

- Click **OK**.

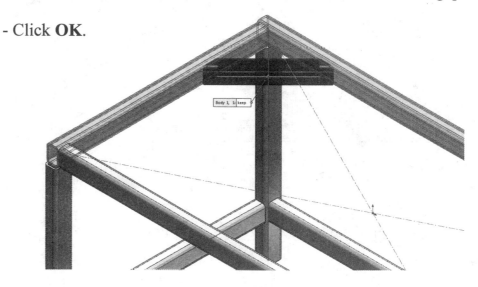

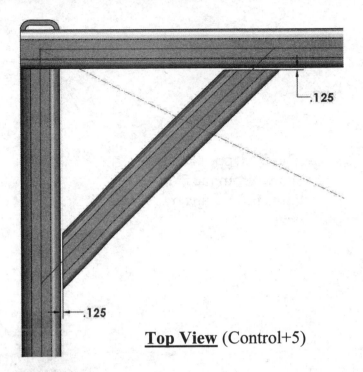

- The first Upper Brace is shown from the Top view with **.125in** gaps on either ends.

Top View (Control+5)

11. Trimming the second Upper Braces:

- Trim the Upper Brace using the **End Trim** and **Bodies** options.

- Enable the **Weld Gap** checkbox and enter: **.125in** for gap.

- Enable **Allow Extension**.

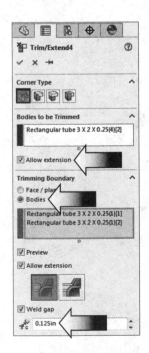

- Click **OK**.

Top View (Control+5)

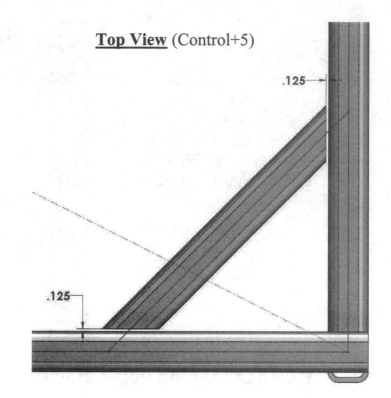

.125

- The 2ⁿᵈ Upper Brace is shown from the Top view with **.125in** gaps on either ends.

.125

12. Updating the Cut list:

- The Cut List should be updated after a new group is added to the weldment part. Identical items are grouped together in Cut-List-Item subfolders.

- ▦ indicates the Cut list is up to date.

- ▦ indicates the Cut list should be updated.

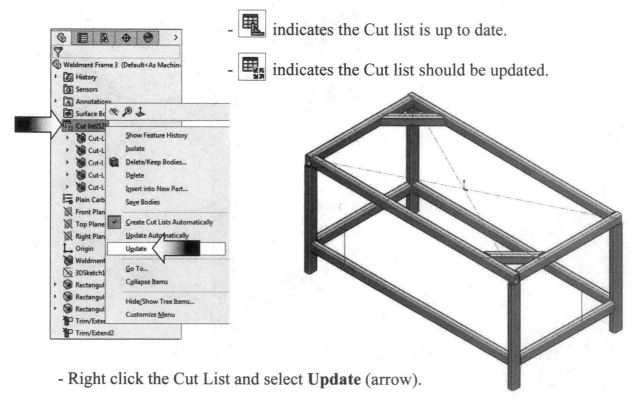

- Right click the Cut List and select **Update** (arrow).

13. Creating the Lower Braces:

- Click the **Structural Member** command 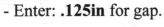.

- Use the <u>same selections</u> as the other 4 groups.

- Select the **2 angled lines** on the <u>bottom</u> of the main sketch.

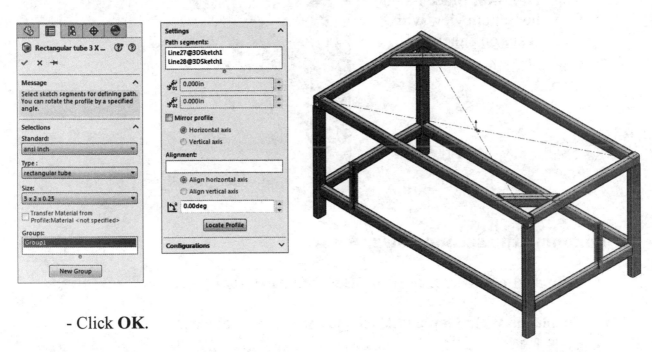

- Click **OK**.

14. Trimming the first Lower Brace:

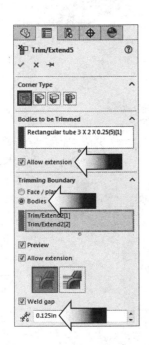

- Trim the Lower Brace using the **End Trim** and **Bodies** options.

- Enable the **Weld Gap** checkbox (arrow).

- Enable the **Allow Extension** checkbox.

- Enter: **.125in** for gap.

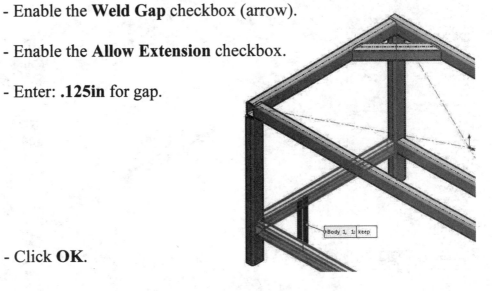

- Click **OK**.

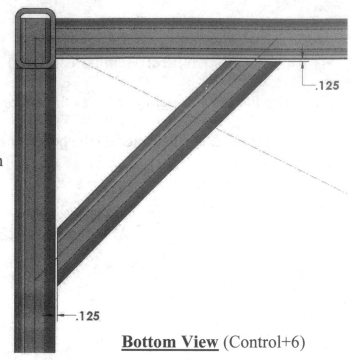

- The first Lower Brace is shown from the Bottom view with **.125in** gaps on either ends.

.125

Bottom View (Control+6)

15. Trimming the second Lower Brace:

- Trim the Lower Brace using the **End Trim** and **Bodies** options.

- Enable the **Weld Gap** checkbox and enter **.125in** for gap.

- Enable **Allow Extension**.

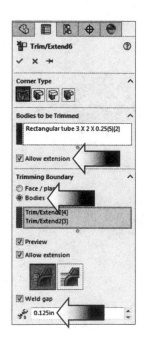

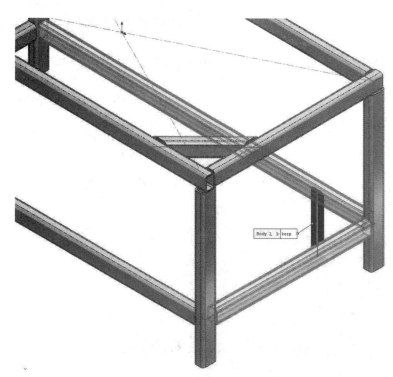

- Click **OK**.

Bottom View (Control+6)

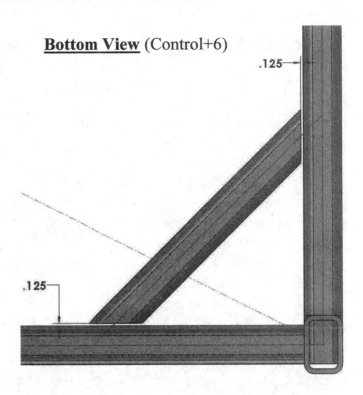

- The second Lower Brace is shown from the Top view with **.125in** gaps on either ends.

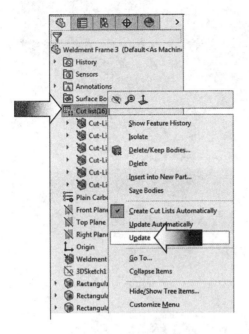

- **Update** the Cut List once again to group the identical structural members into sub-folders.

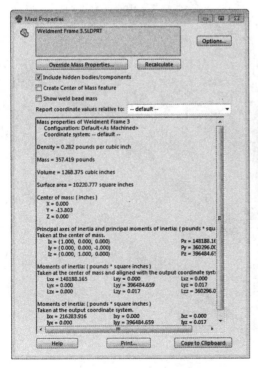

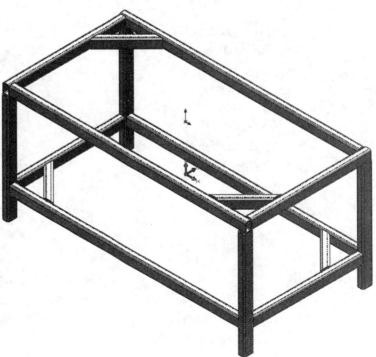

- Enter the mass here: _____ Pounds.

16. Adding End Caps:

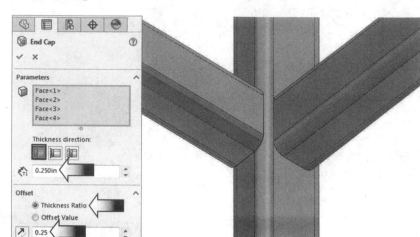

- End caps can be added, including internal end caps to close off open structural members.

- To apply fillets or chamfers to all end caps, under Corner Treatment, select Fillet or Chamfer.

- You can offset an end cap from the inside face of a structural member by specifying a decimal Offset value in addition to the Thickness ratio.

- Click the **End Cap** command.

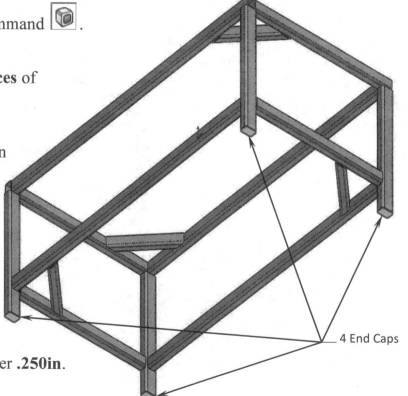

- Select the **4 bottom faces** of the 4 vertical tubes.

- For Thickness Direction click **Outward** and enter **.250in** for thickness.

- For Offset Thickness Ratio, enter **.250in**.

- For Corner Treatment, Click **Chamfer** and enter **.250in**.

4 End Caps

- Click **OK**.

17. Calculating the final mass:

- Select the **Mass Properties** command from the Evaluate tab.

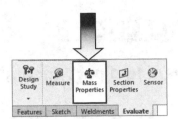

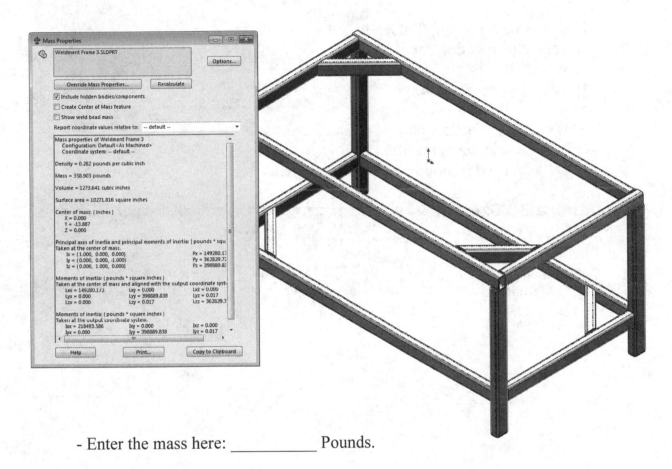

- Enter the mass here: _____ Pounds.

18. Saving your work:

- Select **File / Save As**.

- Enter **Weldment Frame 3** for the name of the file.

- Click **Save** and overwrite the old file with your completed one.

- Close all documents.

CHALLENGE 3

1. Opening a part document:

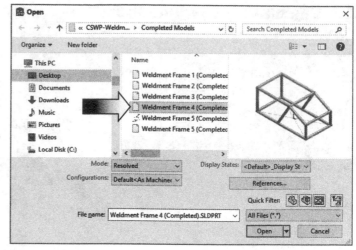

- Select **File / Open**.

- Browse to the Training Folder and open the part document named **Weldment Frame 4.sldprt**.

- This challenge focuses on creating a weldment part from a 3D Sketch and trimming the Structural Members.

2. Creating a 3D Sketch:

- Click the **Weldment** command to create a Weldment feature on the tree.

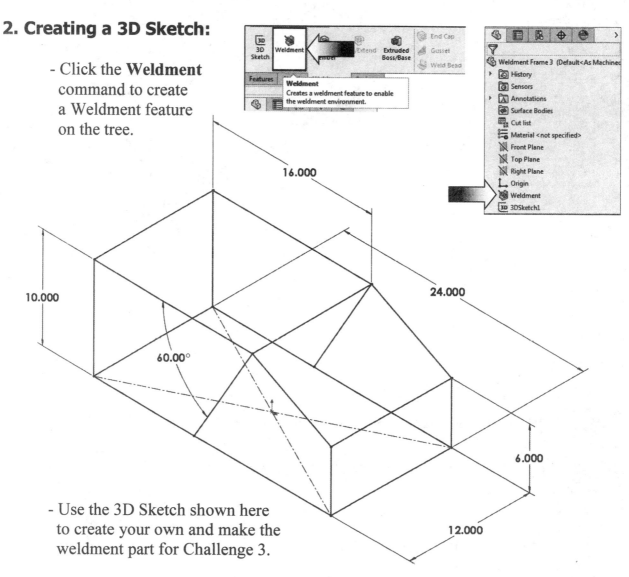

- Use the 3D Sketch shown here to create your own and make the weldment part for Challenge 3.

- Select **3D Sketch** command (arrow) below the
  Sketch drop-down arrow.

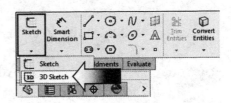

- Select the **Center Rectangle** command, hold
  the **Control** key and click the **Top** plane (arrow).

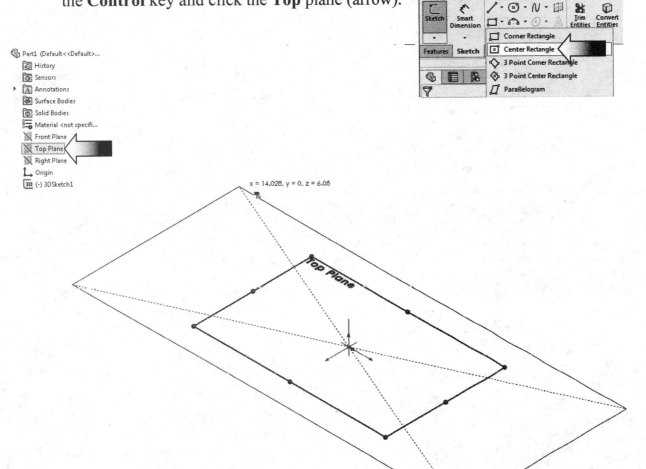

- Sketch a rectangle centered on the origin.

- An alternative way to sketch the same rectangle on the Top plane is to push the
  TAB key once or twice and look out for the **ZX** indicator to appear next to the
  mouse cursor, prior to making the rectangle.

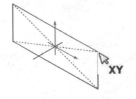
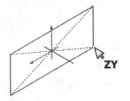
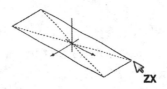

- Add the **Along Z** (Vertical) relation to the line on the left side of the rectangle.

- Add the **Along X** (Horizontal) to the line at the bottom.

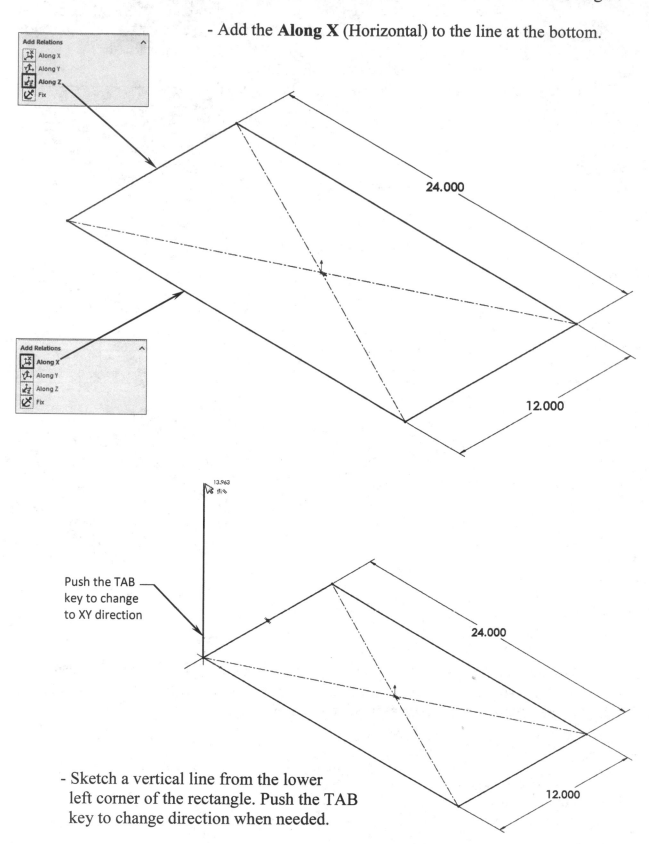

- Sketch a vertical line from the lower
  left corner of the rectangle. Push the TAB
  key to change direction when needed.

- Sketch **3 more vertical lines** and add an **Equal relation** to each pair as noted.

- Also add the height dimensions to the vertical lines.

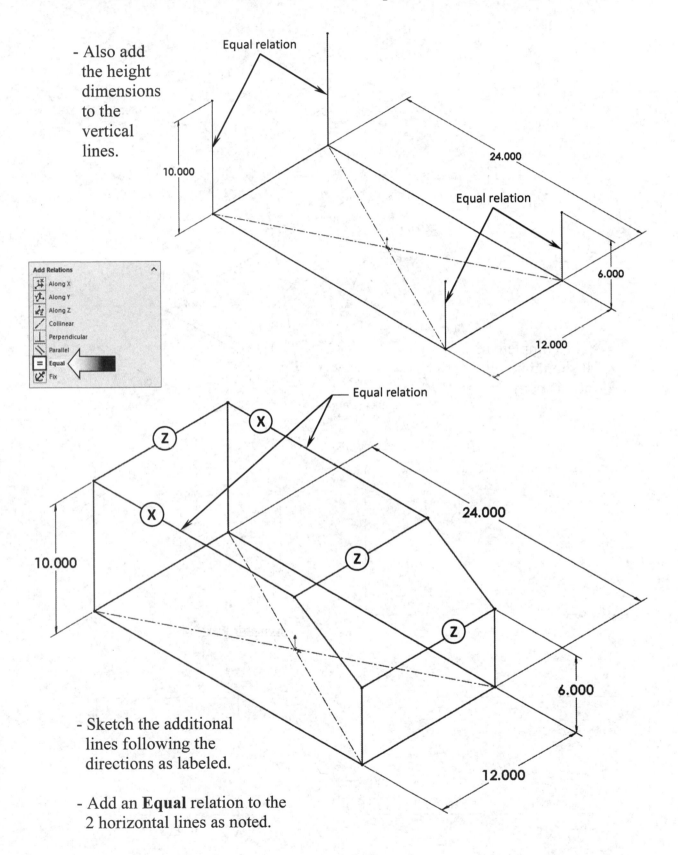

- Sketch the additional lines following the directions as labeled.

- Add an **Equal** relation to the 2 horizontal lines as noted.

- Sketch the **two 60 degree lines** and add a **Parallel** relation to constraint them.

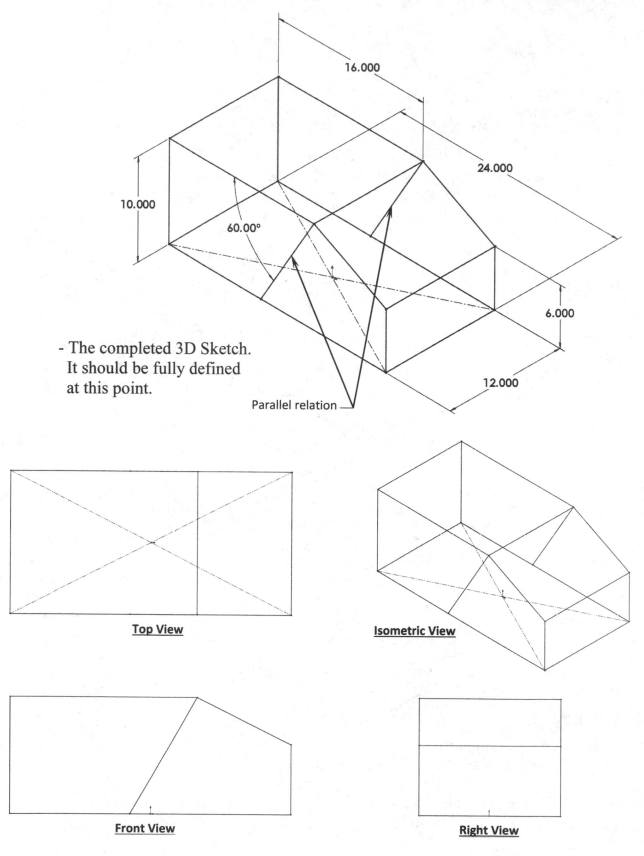

- The completed 3D Sketch.
  It should be fully defined
  at this point.

- **Exit** the 3D Sketch (or press Control+Q).

3. Adding Structural Members:

- Click the **Structural Member** command .

- Select the following: * **Ansi Inch**   * **Pipe**   * **0.5 sch 40**

- Select the **6 lines** on top of the 3D Sketch.

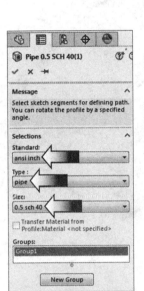 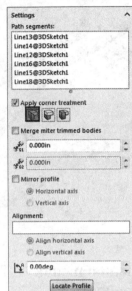 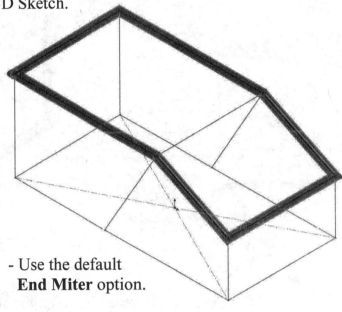

- Use the default
  End Miter option.

- Click **OK**.

- **Repeat** the step 3 and add the same Structural Members to the **4 lines** on the bottom of the 3D Sketch.

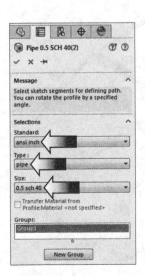  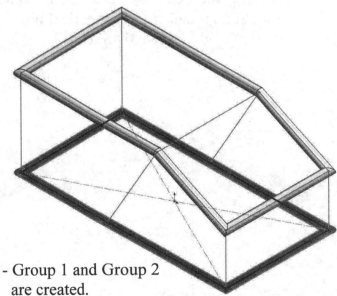

- Group 1 and Group 2
  are created.

4. Creating the Parallel Group:

- Click the **Structural Member** command .

- Use the <u>same selections</u> as the last 2 groups and select the **four vertical lines** from the graphics area.

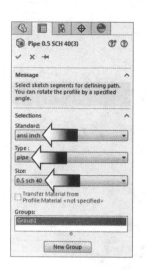  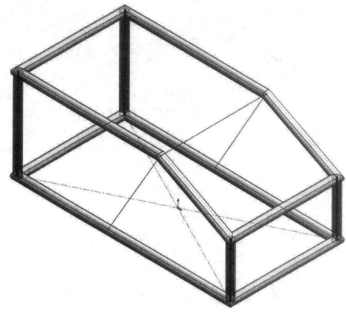

- Click **OK**.

- There is some interference between the structural members. They will be trimmed to the correct lengths in the next couple steps.

- **Repeat** the step above and add the same structural member to the **vertical line** in the middle of the 3D Sketch.

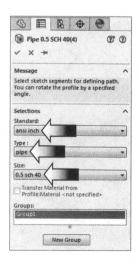 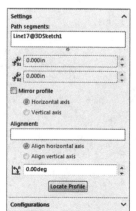 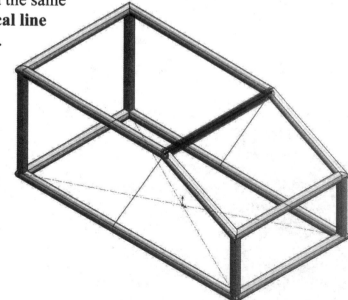

5. Creating the last group:

- Click the **Structural Member** command 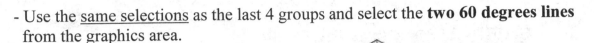.

- Use the <u>same selections</u> as the last 4 groups and select the **two 60 degrees lines** from the graphics area.

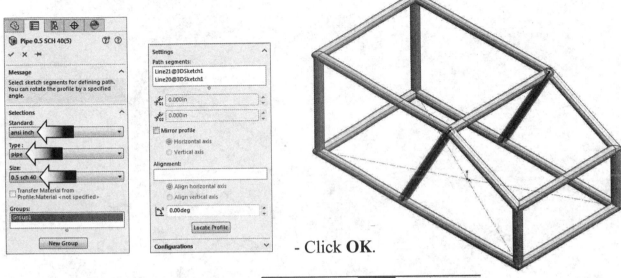

- Click **OK**.

6. Trimming the first tubes:

- Click **Trim/Extend** .

- For Bodies to be Trimmed, select the **vertical tube** as noted and **<u>uncheck</u>** the **Allow Extension** checkbox.

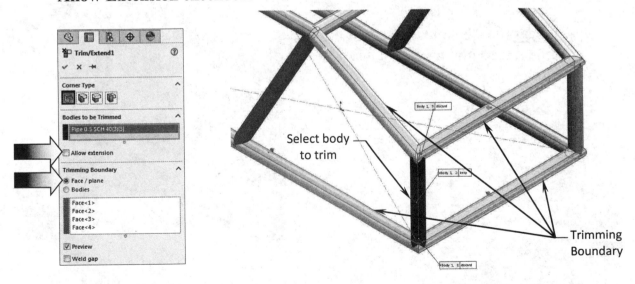

- For Trimming Boundary, select the **Face/Plane** option and select the **4 structural members** as noted.

7. Trimming the second tubes:

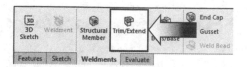

- Click **Trim/Extend** 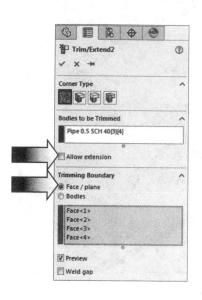.

- For Bodies to be Trimmed, select the **vertical tube** on the far right as noted. **Uncheck** the **Allow Extension** checkbox.

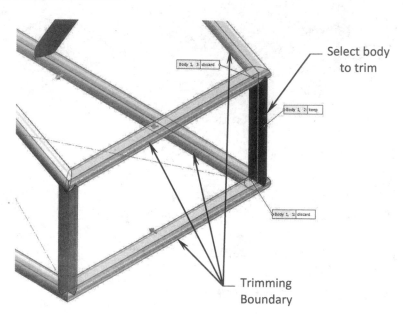

Select body to trim

Trimming Boundary

- For Trimming Boundary, select the **Face/Plane** option and select the **4 structural members** as noted.

8. Trimming the third tubes:

- **Repeat** the step above and trim the vertical tube on the front-left as noted.

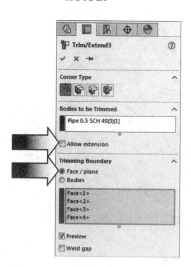

Select body to trim

Trimming Boundary

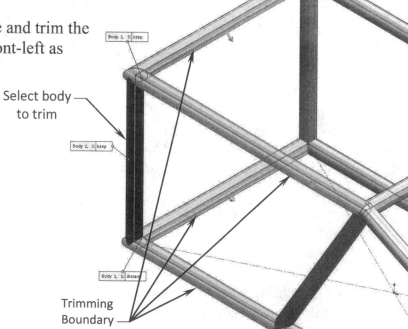

9. Trimming the fourth tubes:

- Click **Trim/Extend**.

- For Bodies to be Trimmed, select the **fourth vertical tube** on rear-left.

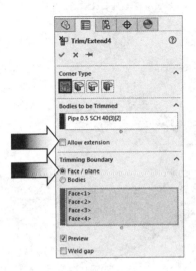

- Clear the **Allow Extension** check box.

- For **Trimming Boundary**, select the Face/Plane option and select the 4 structural members as noted.

- Click **OK**.

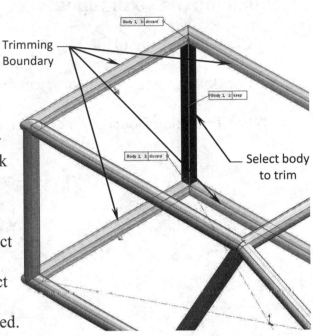

10. Trimming the fifth tubes:

- Click **Trim/Extend**.

- For Bodies to be Trimmed, select the **two 60 degrees tubes**.

- For Trimming Boundary, select the **6 structural members** as indicated.

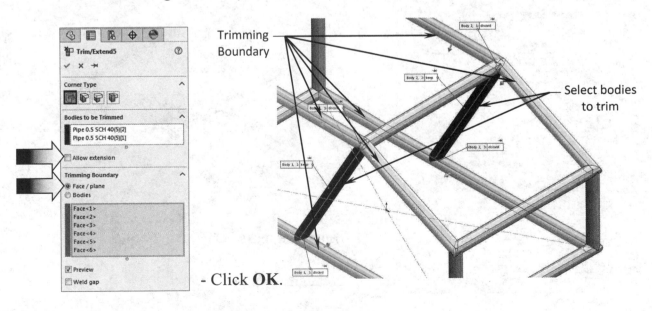

- Click **OK**.

11. Trimming the sixth tubes:

- Click **Trim/Extend** .

- For Bodies to be Trimmed, select the **support tube** in the middle of the weldment.

- For Trimming Boundary, select the **6 structural members** as indicated.

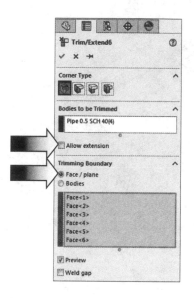

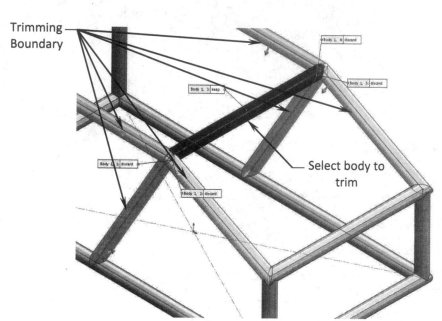

Trimming
Boundary

Select body to
trim

- Click **OK**.

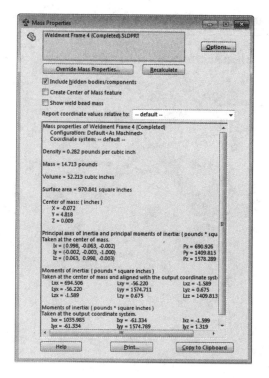

- Assign the material **Plain Carbon Steel** to the weldment part.

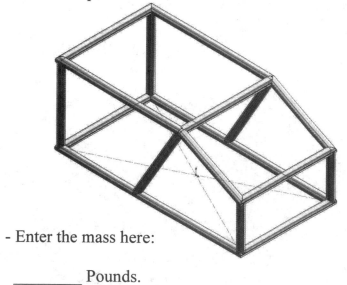

- Enter the mass here:

_____ Pounds.

CHALLENGE 4

1. Opening a part document:

- Select **File / Open**.

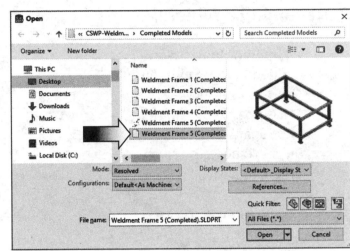

- Browse to the Training Folder and open the part document named **Weldment Frame 5.sldprt**.

- This challenge focuses on Cut List Management, and Cut List Creation in a Weldment Drawing.

2. Assigning material:

- Right click the Material option and select **Plain Carbon Steel**.

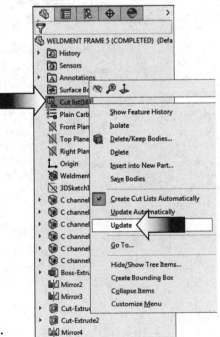

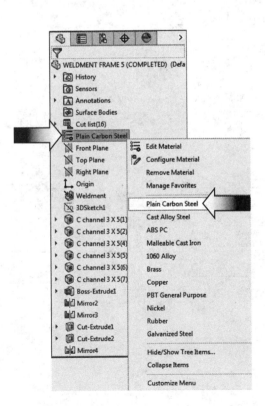

- Right click the Cut List and select **Update**. The identical members will be grouped into subfolders.

3. Renaming a folder:

- The Foot Plates were added to the weldment part as extruded features. They should be renamed to make it more descriptive on the Cut List.

- Expand the Cut List and <u>Rename</u> the folder to **Foot Plates** (arrow).

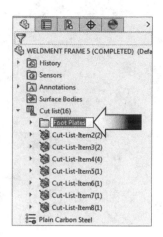
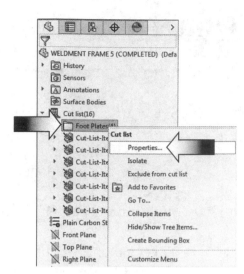

4. Modifying the Properties:

- Right click the Foot Plate folder and select **Properties**.

- Click in line item 3 and enter **DESCRIPTION** (arrow). Enter **FOOT PLATES** under Value / Text Expression.

- In line item 4, enter **LENGTH** (arrow). Enter: **5.00 X 3.00**.

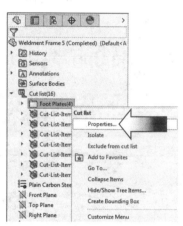

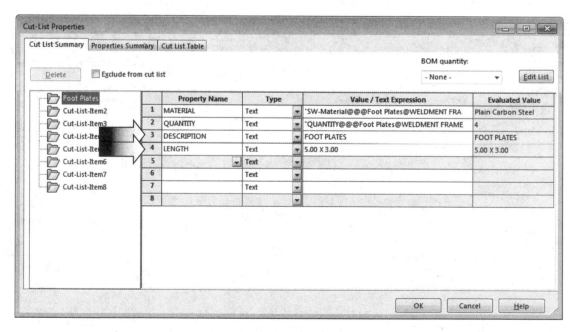

		Property Name	Type	Value / Text Expression	Evaluated Value
Foot Plates	1	MATERIAL	Text	"SW-Material@@@Foot Plates@WELDMENT FRA	Plain Carbon Steel
Cut-List-Item2	2	QUANTITY	Text	"QUANTITY@@@Foot Plates@WELDMENT FRAME	4
Cut-List-Item3	3	DESCRIPTION	Text	FOOT PLATES	FOOT PLATES
Cut-List-Ite	4	LENGTH	Text	5.00 X 3.00	5.00 X 3.00
Cut-List-Ite	5		Text		
Cut-List-Item6	6		Text		
Cut-List-Item7	7		Text		
Cut-List-Item8	8		Text		

- Click **OK**.

5. Transferring to a drawing:

- Select **File / Make Drawing From Part** (arrow).

- Select the default **Drawing Template** and click **OK**.

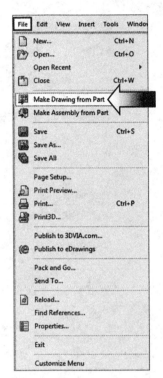

- For this challenge, use the default **A-Size** paper.

- Set the Projection to **3rd Angle**.

- Set Drafting Standards to **ANSI**.

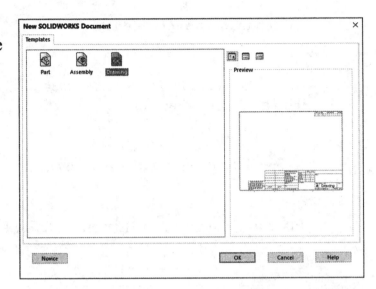

6. Adding a drawing view:

- Expand the **View Palette** (arrow).

- Drag and drop the **Isometric view** to the drawing approximately as shown.

- Change the view **Scale** to **1:24**.

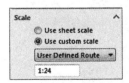

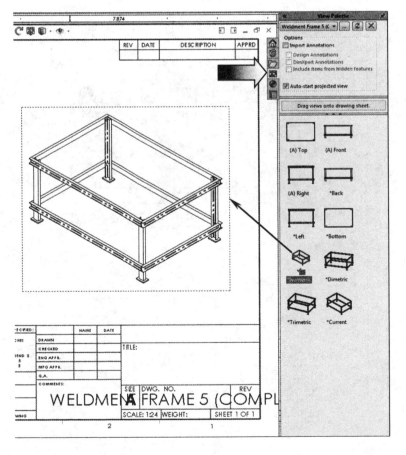

7. Inserting the Weldment Cut List:

- Click the dotted border of the Isometric view to activate it.

- Switch to the **Annotation** tab.

- Select the **Weldment Cut List** under the Tables drop-down (arrow).

- Keep all the default settings as they are and click **OK**.

- Place the Weldment Cut List on the left side of the isometric view.

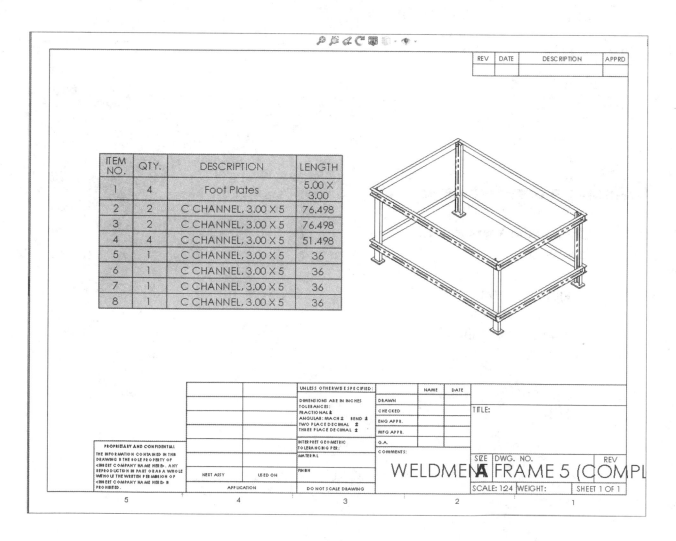

ITEM NO.	QTY.	DESCRIPTION	LENGTH
1	4	Foot Plates	5.00 X 3.00
2	2	C CHANNEL, 3.00 X 5	76.498
3	2	C CHANNEL, 3.00 X 5	76.498
4	4	C CHANNEL, 3.00 X 5	51.498
5	1	C CHANNEL, 3.00 X 5	36
6	1	C CHANNEL, 3.00 X 5	36
7	1	C CHANNEL, 3.00 X 5	36
8	1	C CHANNEL, 3.00 X 5	36

8. Adding Balloons:

- Click **Auto Balloons** 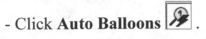.

- **Clear** Insert Magnetic Lines and change the Balloon Settings to **1 Character**.

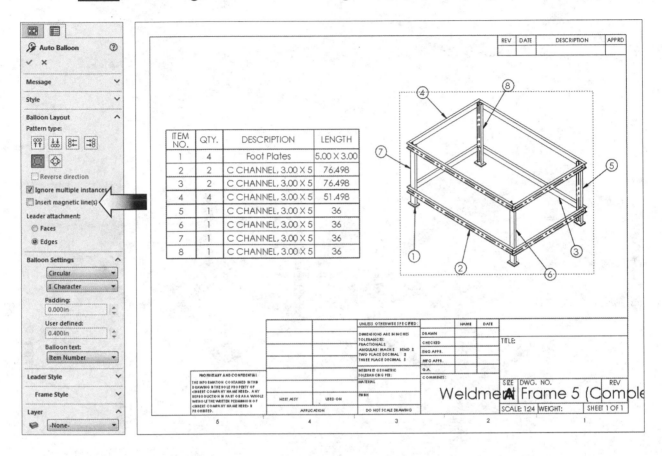

- Click **OK** and re-arrange the balloons to look similar to the one displayed above.

- Locate **Item No. 4**,
Find the Length of the
C Channel, 3.00 X 5
and enter it here:

_____

- **Save** your work as
Weldment Frame 5.slddrw.

- Close all documents.

ITEM NO.	QTY.	DESCRIPTION	LENGTH
1	4	Foot Plates	5.00 X 3.00
2	2	C CHANNEL, 3.00 X 5	76.498
3	2	C CHANNEL, 3.00 X 5	76.498
4	4	C CHANNEL, 3.00 X 5	51.498
5	1	C CHANNEL, 3.00 X 5	36
6	1	C CHANNEL, 3.00 X 5	36
7	1	C CHANNEL, 3.00 X 5	36
8	1	C CHANNEL, 3.00 X 5	36

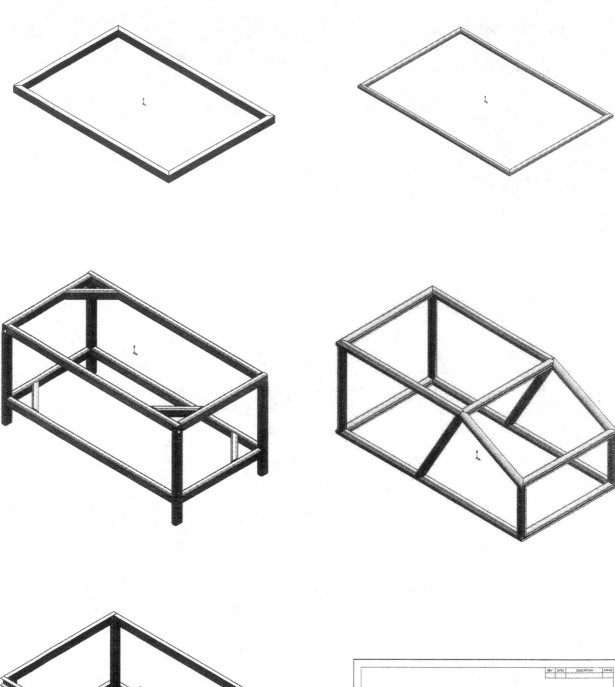

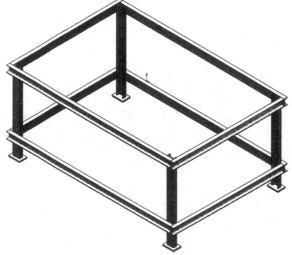

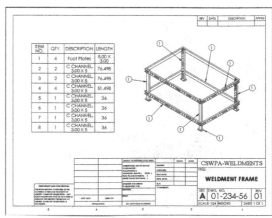

CHAPTER 4

Sheet Metal

CSWPA – Sheet Metal

The completion of the Certified SOLIDWORKS Professional Advanced Sheet Metal (CSWPA-SM) exam shows that you have successfully demonstrated your ability to use SOLIDWORKS Sheet Metal tools.

Employers can be confident that an individual with this certification understands the set of SOLIDWORKS tools that aid in the design of sheet metal components. Recommended training courses: SOLIDWORKS Essentials and Sheet Metal.

Note: A new version of the Sheet Metal exam in English has been released and requires the use of SOLIDWORKS 2011 or later. For now, all non-English versions of the test will still be the older Sheet Metal exam (2006 - 2010 version).

Exam Length: 2011 Version (English): 90 minutes
                    2006 - 2010 Version (Non-English): 2 hours

Minimum Passing grade: 75%

Re-test Policy: There is a minimum 30 day waiting period between every attempt of the CSWPA-SM exam. Also, a CSWPA-SM exam credit must be purchased for each exam attempt. All candidates receive electronic certificates and a personal listing on the CSWP directory when they pass.

Exam features hands-on challenges in many of these areas of SOLIDWORKS Sheet Metal functionality such as:

> Create flanges, closed Corners, Gauge Tables, Bend Calculation Options, Hem, Jogs, Sketched Bends, Fold and Unfold, Forming Tools, Flatten, Convert to Sheet Metal, and Sheet metal Cut List Properties.

Excel should be installed on the computer you are running SOLIDWORKS to ensure that your Gauge Table functionality works properly for the exam.

CSWPA – Sheet Metal

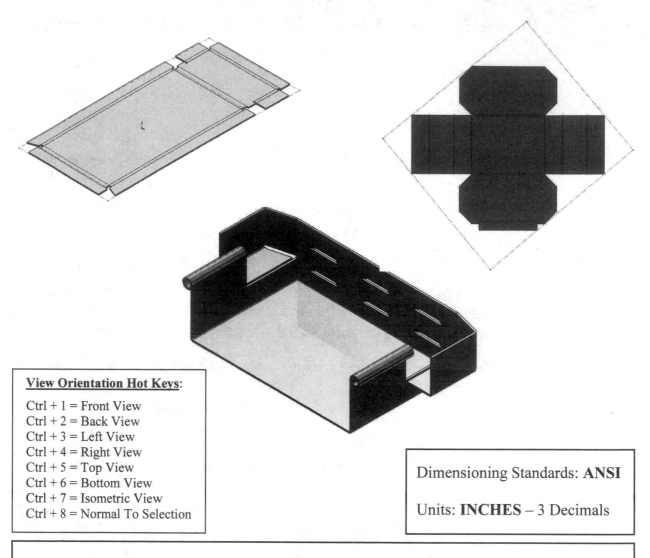

Dimensioning Standards: **ANSI**

Units: **INCHES** – 3 Decimals

Tools Needed:

 Convert to Sheet Metal

 Base Flange

 Hem

 Edge Flange

 Miter Flange

 Mirror

 Linear Pattern

 Forming Tool

 Cut List

CHALLENGE 1

1. Opening a part document:

- Select **File / Open**.

- Browse to the Training Folder and open the part document named: **Convert to SM.sldprt**.

- This challenge examines your skills on converting a solid model into a sheet metal model.

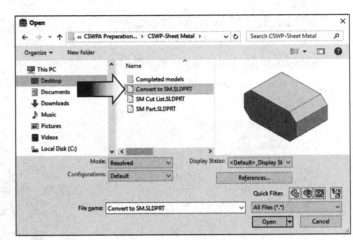

2. Enabling the Sheet Metal tab:

- Right click the **Evaluate** tab and enable the **Sheet Metal** tab (arrow).

- Switch to the **Sheet Metal** tab.

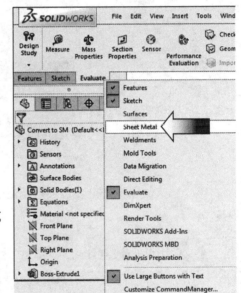

- We will be creating a sheet metal part that fits around the existing model.

- A copy is made from the original model and then gets converted into sheet metal part.

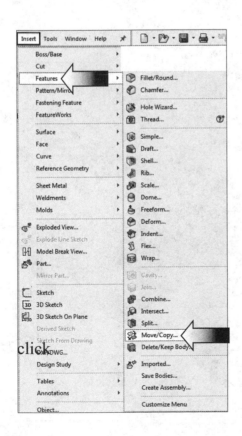

3. Making a copy of the solid model:

- Select **Insert / Features / Move/Copy** (arrow).

- If the Move/Copy options are not available, 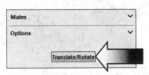 the **Translate/Rotate** button at the bottom of the tree to change to the **Move/Copy** options.

click

- For Bodies to Move/Copy, <u>select the model</u> in the graphics area.

- Enable the **Copy** checkbox (arrow).

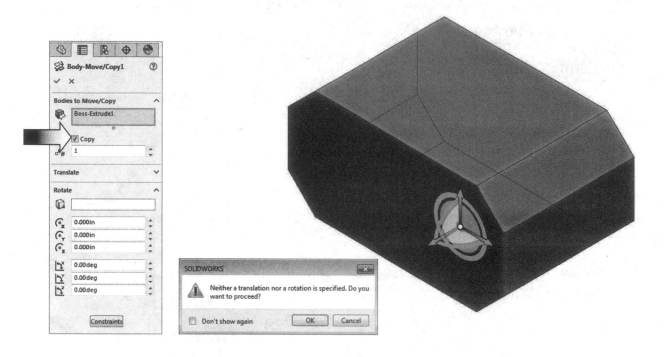

- Leave all distances at their **Zero** values.

- Click **OK**. Click OK again to proceed with the move.

- A copy of the solid model is created and placed on top of the original body.

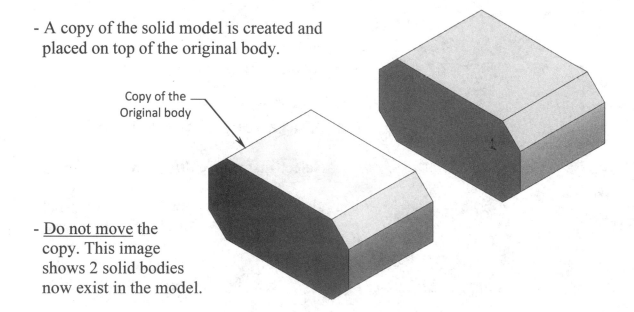

Copy of the
Original body

- <u>Do not move</u> the copy. This image shows 2 solid bodies now exist in the model.

4. Converting to Sheet Metal:

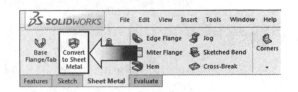

- Click **Convert to Sheet Metal**.

- Select the **Fixed Face** as noted.

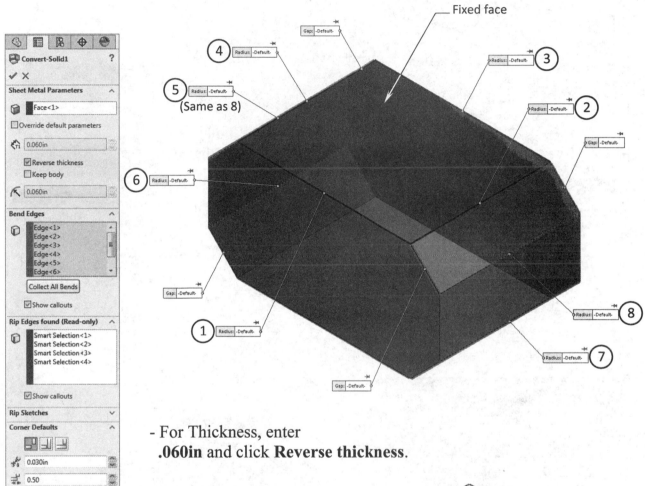

Fixed face

- For Thickness, enter
  .060in and click **Reverse thickness**.

- For Bend Edges, select
  the **8 edges** as indicated.

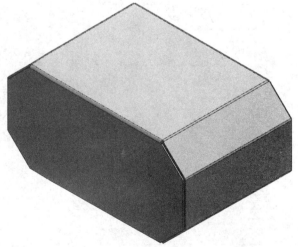

- Enter **.030in**. for Gap.

- Leave Overlap at the default **0.50**

- Set Auto Relief to **Tear, 0.50**

- Click **OK**.

5. Calculating the mass:

- Change the Material to **1060 Alloy**.

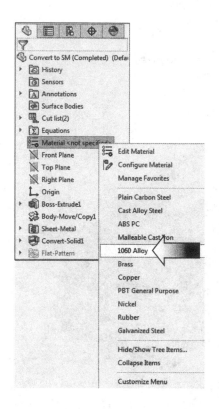

- Switch to the **Evaluate** tab.

- Click **Mass Properties** 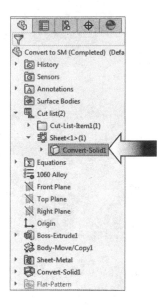.

- Select the new **solid body** (Convert Solid1) under the Cut List folder.

- Enter the mass of the converted part here:

_____ Pounds.

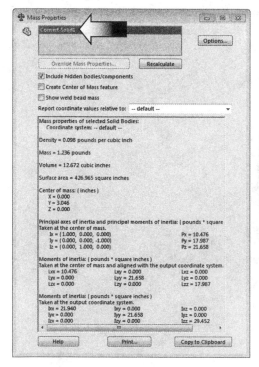

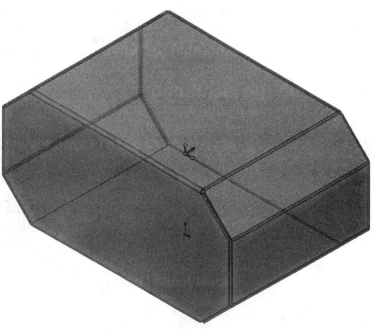

6. Measuring the Flat Pattern:

- Right click the converted model and select **Toggle Flat Display** (arrow).

- This option "superimposes" the flat pattern over the folded for viewing purposes.

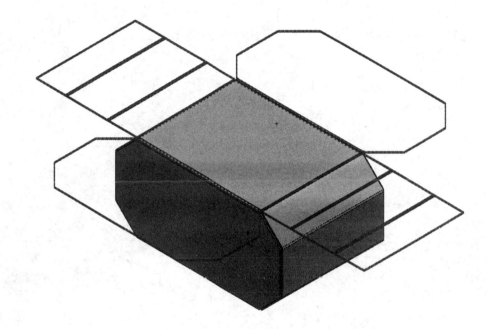

- Press **Esc** to cancel the Toggle Flat Pattern mode.

- Click the **Flatten** command 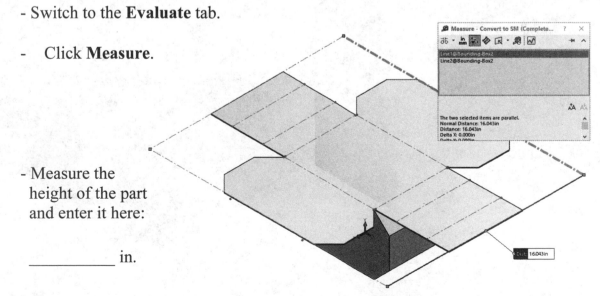 on the Sheet Metal tab to flatten the part.

- Switch to the **Evaluate** tab.

- Click **Measure**.

- Measure the height of the part and enter it here:

_____ in.

7. Creating a Hem:

- Turn off the Flatten command and click **Hem** .

- Select the **edge** as indicated and click the **Rolled** type (arrow).

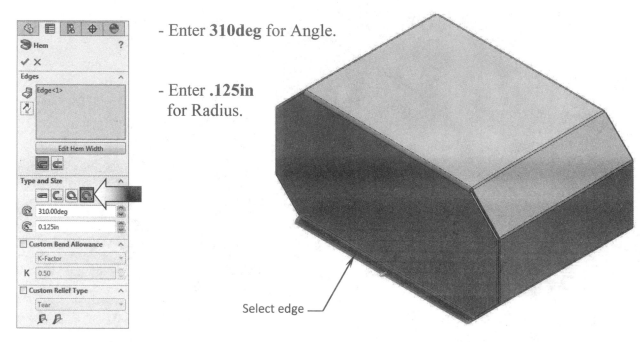

- Enter **310deg** for Angle.

- Enter **.125in** for Radius.

Select edge ⎯

- Click **Edit Hem Width** (arrow).

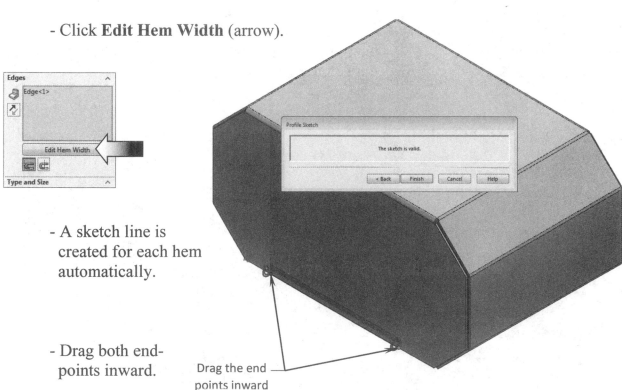

- A sketch line is created for each hem automatically.

- Drag both end-points inward.

Drag the end ⎯ points inward

- The line itself has an On-Edge relation with the edge of the model, but the ends should be controlled with distance dimensions.

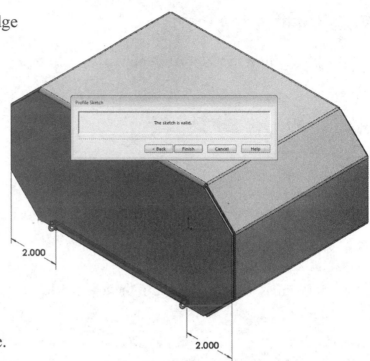

- Add a distance of **2.000in** to each end of the line as shown.

- Attach the dimensions to the outer edges of the flange.

- Click **Finish**.

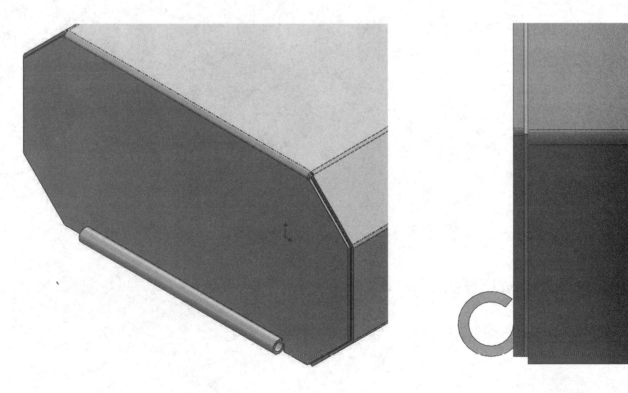

Isometric View   Right View

8. Calculating the mass:

- Switch to the **Evaluate** tab.

- Click **Mass Properties** .

NOTE: Is it important when selecting the solid body to calculate the mass. Be sure to select the converted body every time.

- Expand the **Cut List** and the **Sheet Metal** folders.

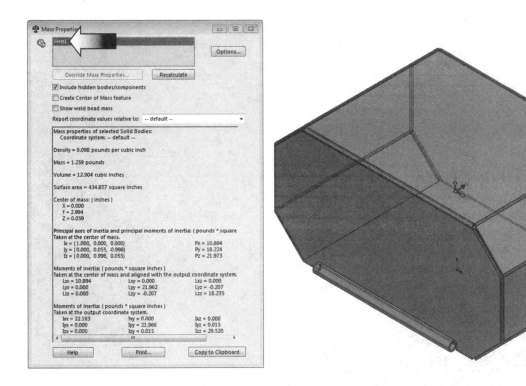

- Select the converted model (**Hem1**) and enter the mass here:

_____ Pounds.

- **Save** and close all documents.

CHALLENGE 2

1. Opening a part document:

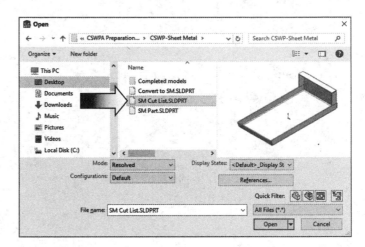

- Select **File / Open**.

- Browse to the Training Folder
  and open the part document
  named **SM Cut list.sldprt**.

- This challenge examines your
  skills on modifying the bend
  radiuses and retrieving the Cut
  List Properties.

2. Modifying the first bend radius:

- This model has several flanges but we will only change the radiuses of the Edge-
  Flange1 and Edge Flange2.

- Locate the **Edge Flange1** from the FeatureManager.
  Right click it and select **Edit Feature** (arrow).

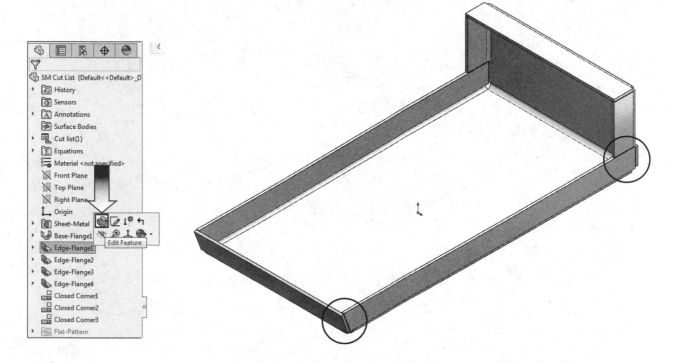

- Enable the **Use Gauge Table** checkbox.

- Change the default bend radius to **.125in** (arrow).

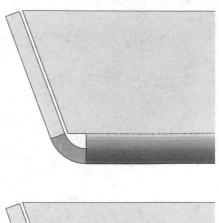

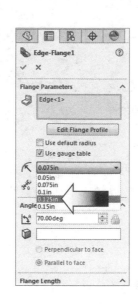

- Keep all other parameters at their default values.

- Click **OK**.

3. Measuring the flat length:

- Switch to the **Evaluate** tab.

- Click **Measure** 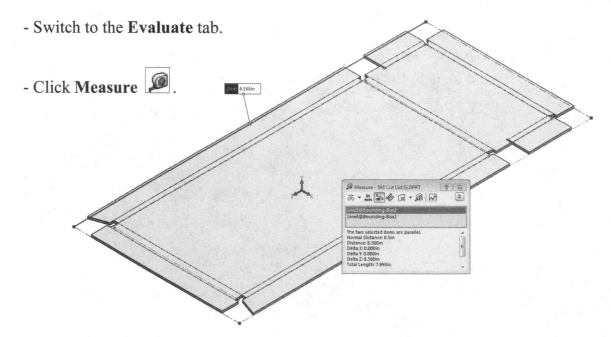.

- Measure the length of the flat pattern (the longest length) and enter it here:

_____ in.

4. Modifying the 2ⁿᵈ bend radius:

- Locate the **Edge Flange2** from the FeatureManager.
  Right click it and select **Edit Feature** (arrow).

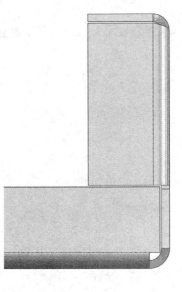

- Enable the **Use Gauge Table** checkbox.

- Change the bend radius to **.125in**

- Keep all other parameters at their default values.

- Click **OK**.

5. Measuring the flat length:

- Click **Measure**.

- Measure the length
  of the flat pattern
  (the longest length)
  and enter it here:

  _____ in.

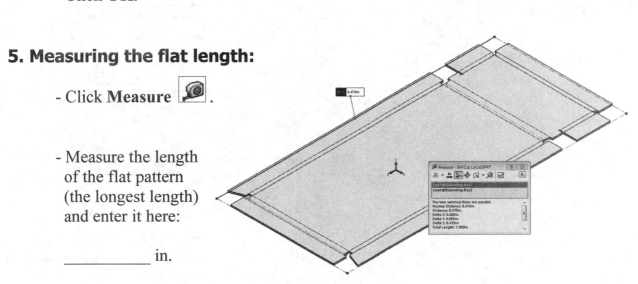

6. Finding the Bounding Box Area:

- Right click the **Cut list** and select **Update** (arrow).

- Expand the **Cut List**.

- Right click the item **Sheet<1>** and select **Properties**.

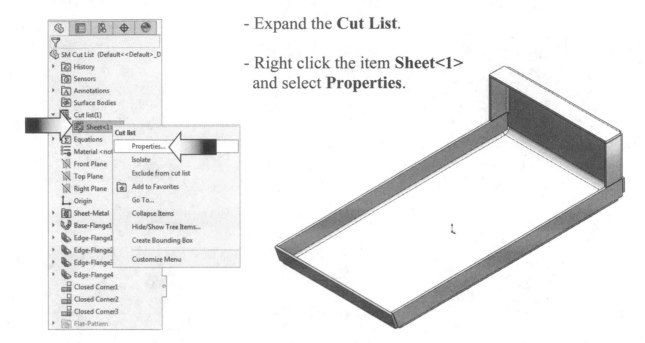

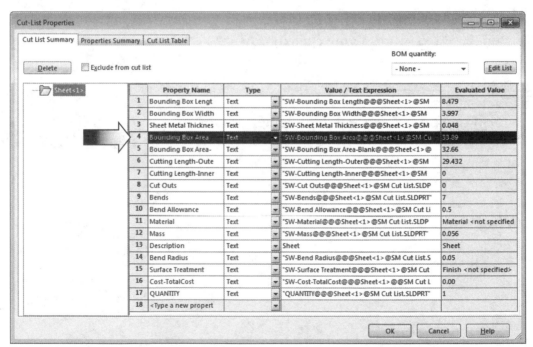

- Locate the row number 4 and enter the **Bounding Box Area** Evaluated Value

here: _____ in^2.   - **Save** and close all documents.

CHALLENGE 3

1. Opening a part document:

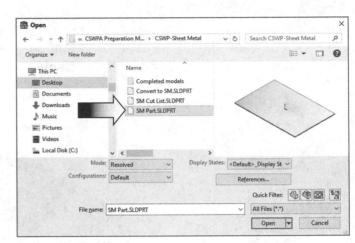

- Select **File / Open**.

- Browse to the Training Folder and open the part document named **SM Part.sldprt**.

- This challenge reinforces your skills of creating a sheet metal part and using the forming tools.

2. Creating the Edge Flanges:

- Click the **Edge Flange** command.

- Select the **3 edges** of the Base Flange as noted.

- Set the following: * **Gap: .010in**   * **Blind: 2.000in**

                         * **Outer Virtual Sharp**   * **Bend Outside**

- Click **OK**.

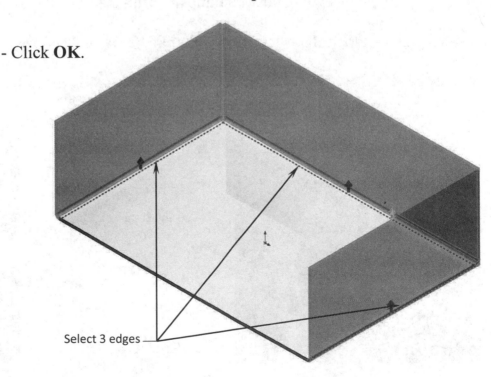

Select 3 edges

3. Closing the corners:

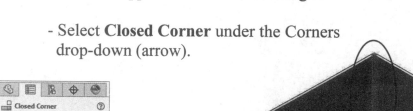

- The back flange needs to be extended and overlapped with the 2 side flanges.

- Select **Closed Corner** under the Corners drop-down (arrow).

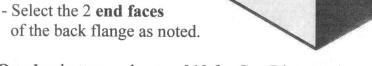

- Select the 2 **end faces** of the back flange as noted.

- Select the **Overlap** button and enter **.010** for Gap Distance (arrow).

- Keep the Overlap/Underlap Ratio at **1** (full material thickness).

- Enable **Open Bend Region**, **Coplanar Faces**, and **Narrow Corner** checkboxes.

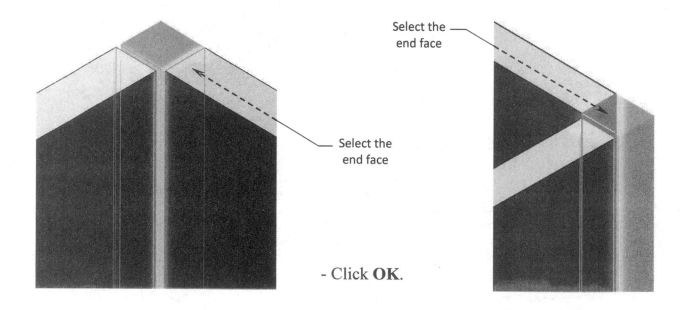

Select the end face

Select the end face

- Click **OK**.

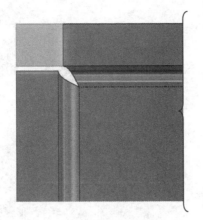 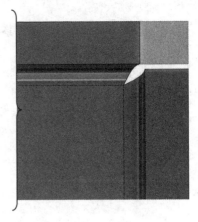

- Change to the Top View (Control+5) and zoom in on the 2 corners to verify the Closed Corner features.

4. Creating a cutout:

- Select the <u>sketch face</u> as noted and open a **new sketch**.

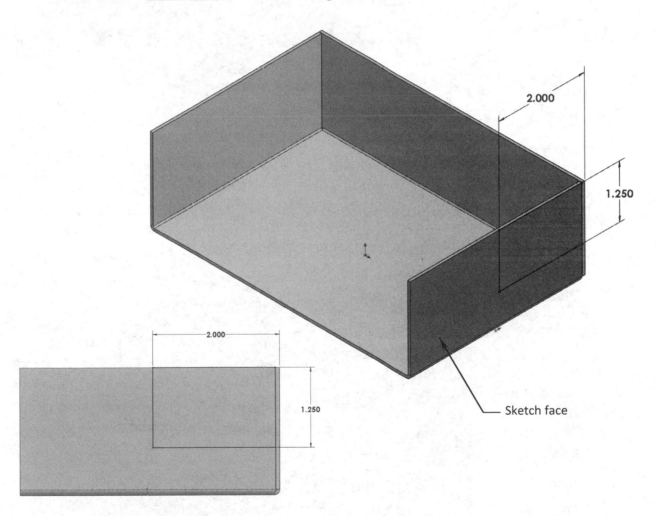

Sketch face

- Sketch a **Rectangle** and add the height and width dimensions as shown. (The dimension 2.00in is measured to the outer edge of the back flange.)

- Switch to the **Features** tab and click **Extruded Cut** 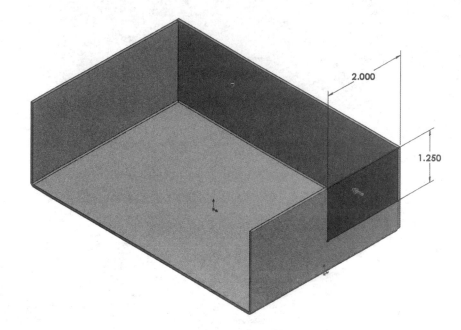.

- Use the default **Blind** cut and enable the **Link to Thickness** checkbox.

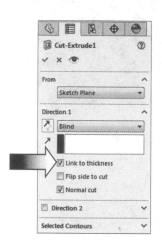

- Click **OK**.

5. Adding a Miter Flange:

- Select the <u>top face</u> of the back flange and open a **new sketch**.

- Sketch **2 lines** from the upper right corner of the flange.

- Add the dimensions shown to fully define the sketch.

- Click the **Miter Flange** command.

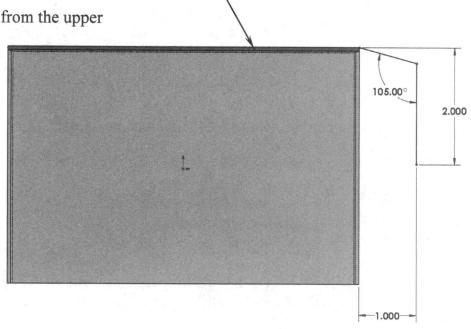

Sketch face

105.00°

2.000

1.000

- Set the following to define the Miter Flange:

* Use Default Radius   *** Flange Position: Material Inside**

* Gap Distance: .010in   *** End Offset: .720in**

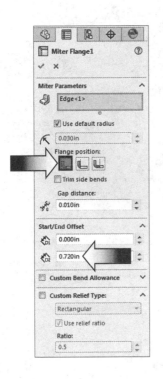

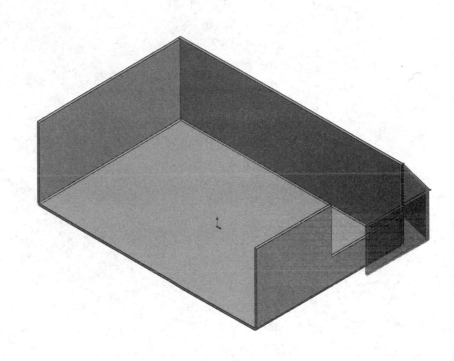

NOTE: *The dimension End Offset .720in creates the required gap **.060in** for this challenge. This gap must remain the same from beginning to end.*

- Change to the Right
  view orientation
  (Control+4).

- Select the **Measure**
  tool to verify the gap.

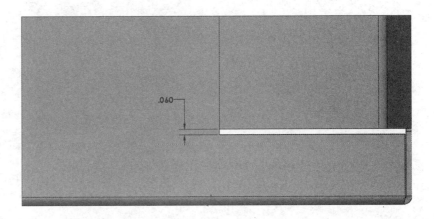

6. Adding a Miter Flange:

- Select the <u>edge</u> as noted and click the **Edge Flange** command.

- Move the cursor towards the left side and click the mouse to lock the direction.

- Select the **Bend Outside** option under the Flange Position section.

- Click the **Edit-Flange Profile** button to edit the sketch.

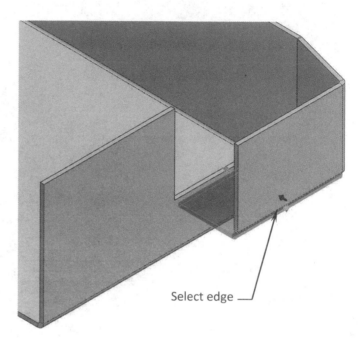

Select edge

- Create an offset of **.060in** from the 2 edges indicated in the image.

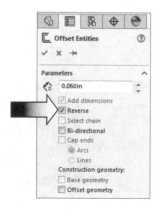

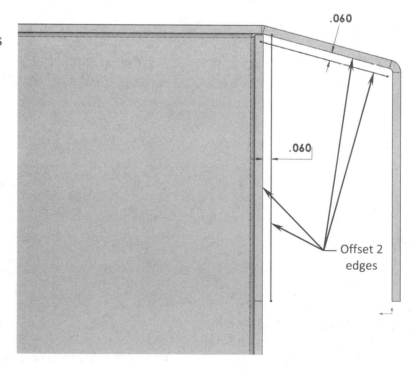

.060

.060

Offset 2 edges

- Click **Reverse** offset to place the lines inside.

- Sketch **2** additional **Lines** as noted. (Lock the right ends of the lines to the existing vertices with the coincident relations.)

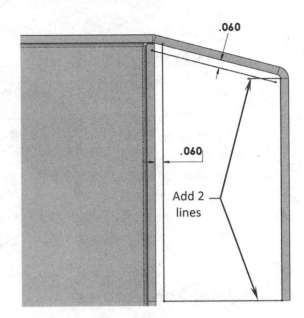

- Trim the overlaps to create a closed profile.

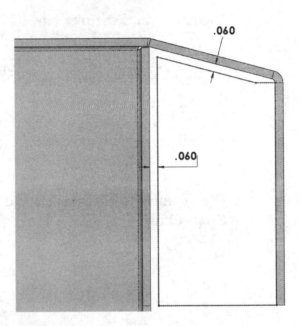

- Change to the Front orientation (Control+1) and verify the gap between the new flange and the part.

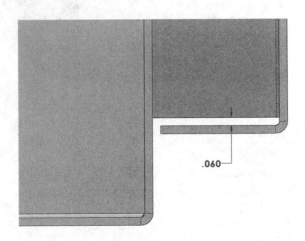

- Click **Finish**.

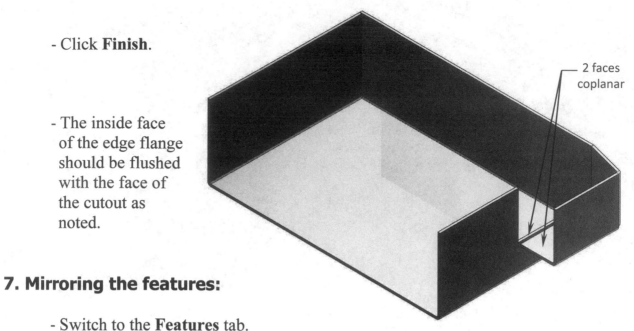

- The inside face of the edge flange should be flushed with the face of the cutout as noted.

2 faces coplanar

7. Mirroring the features:

- Switch to the **Features** tab.

- Click the **Mirror** command .

- For Mirror Face/Plane, select the **Right** plane from the Feature tree.

- For Features to Mirror, select the **Cut-Extrude1**, the **Miter Flange1**, and the **Edge-Flange1**.

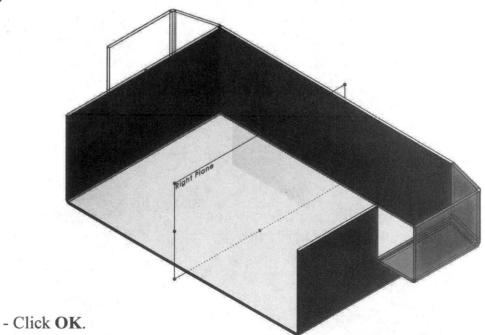

- Click **OK**.

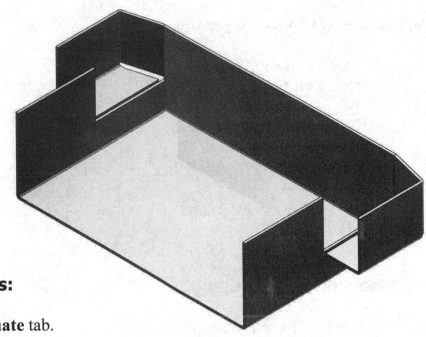

- The features are mirrored about the Right plane.

- The material for this model has been selected as 1060 Alloy.

8. Calculating the mass:

- Switch to the **Evaluate** tab.

- Click **Mass Properties** 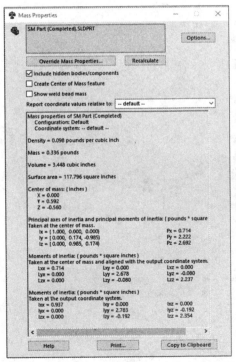 .

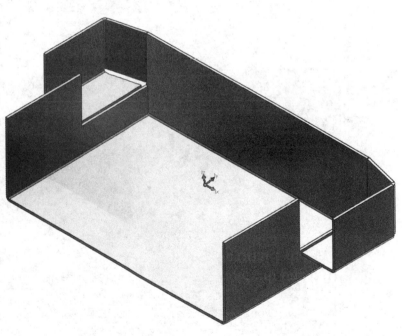

- Locate the mass of the part and enter it here:

_____ Pounds.

9. Changing the K-Factor value:

- Click the **Sheet-Metal** folder and select **Edit Feature** (arrow).

- Change the **K-Factor** to **.48**

- Leave other parameters at their default values.

- Click **OK**.

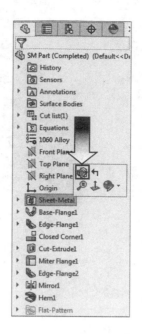

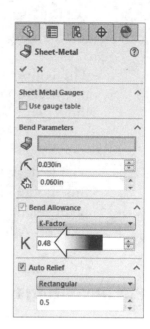

10. Measuring the flat length:

- Select the **Measure** command 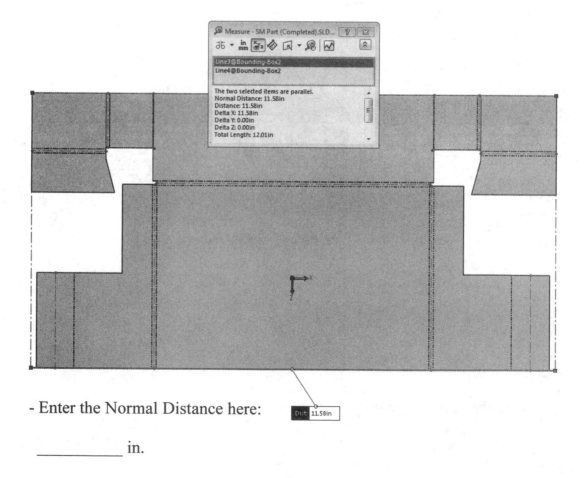 from the **Evaluate** tool tab.

- Measure the overall width dimension (the longest length) of the part.

- Enter the Normal Distance here:

_____ in.

11. Adding Hems:

- Click the **Hem** command .

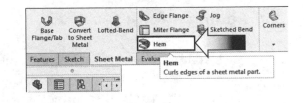

- Select the **2 outer edges** as indicated.

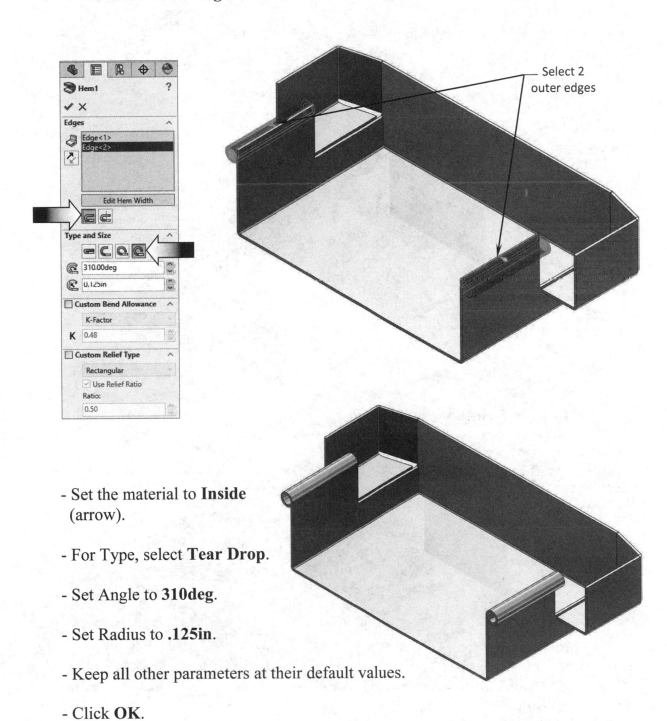

- Set the material to **Inside** (arrow).

- For Type, select **Tear Drop**.

- Set Angle to **310deg**.

- Set Radius to **.125in**.

- Keep all other parameters at their default values.

- Click **OK**.

12. Adding a form feature:

- Expand the **Design Library**, the **Forming Tools**, and the **Louvers folders**.

- Drag and drop the **Louver** form tool to the <u>inside face</u> of the part as shown.

- Change the Rotation Angle to **270deg**.

- Click **Flip Tool** to flip to louver outward.

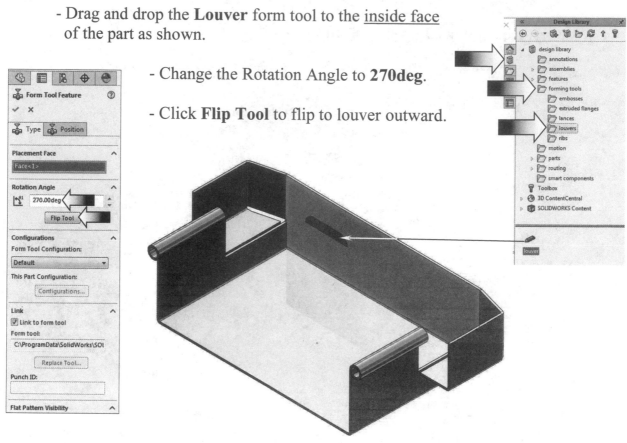

- Click the **Position** tab (arrow).

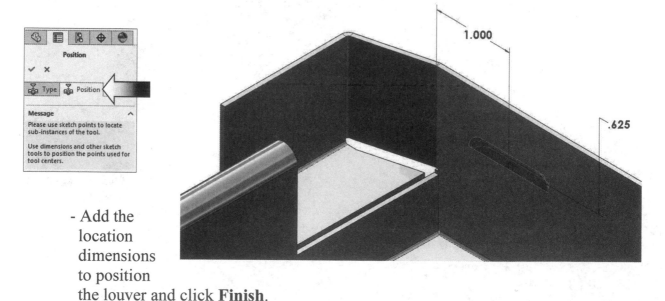

- Add the location dimensions to position the louver and click **Finish**.

13. Creating a linear pattern:

- Switch to the **Features** tab and select the Linear Pattern command (arrow).

- For <u>Direction 1</u>, select the **edge** as indicated.

- For Distance 1, enter **2.000in**.

- For Instances enter **3**.

- For <u>Direction 2</u>, select the **edge** as noted.

- For Distance 2, enter **.750in**.

- For Instances, enter **2**.

- Click **OK**.

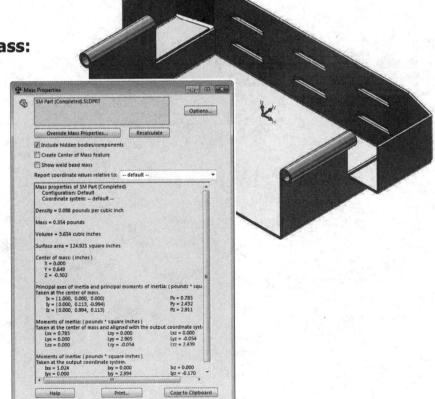

14. Calculating the mass:

- Switch to the **Evaluate** tab.

- Click **Mass Properties** 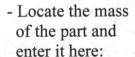.

- Locate the mass of the part and enter it here:

_____ Pounds.

- **<u>Save</u>** and close.

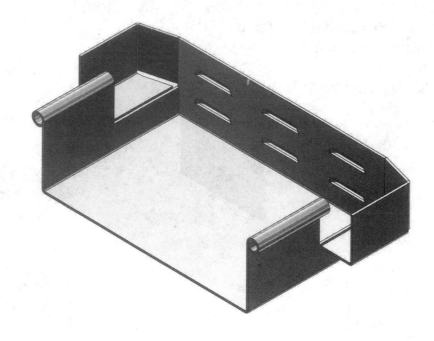

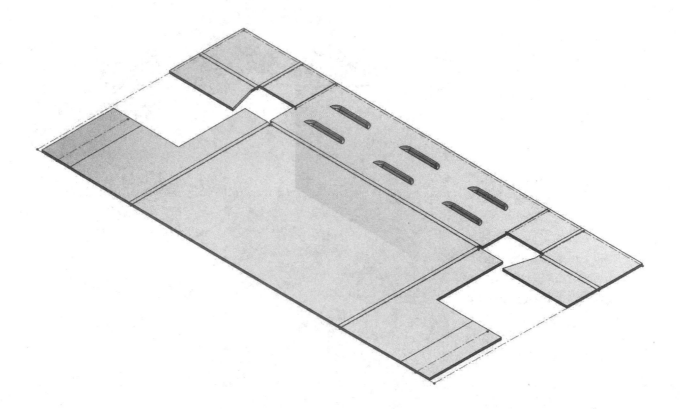

CHAPTER 5

Surfacing

CSWPA – Surfacing

The completion of the Certified SOLIDWORKS Professional Advanced Surfacing (CSWPA-SU) exam shows that you have successfully demonstrated your ability to use SOLIDWORKS Advanced Surfacing tools.

Successful completion of this exam also shows the ability to create advanced surface models as well as troubleshoot and fix broken surface bodies or imported bodies that are incorrect using advanced surfacing techniques.

Recommended Training Courses: Advanced Surface Modeling

Note: You must use at least SOLIDWORKS 2009 for this exam. Any use of a previous version will result in the inability to open some of the testing files.

Exam Length: 90 minutes
Minimum Passing grade: 75%
Re-test Policy: There is a minimum 30 day waiting period between every attempt of the CSWPA-SU exam. Also, a CSWPA-SU exam credit must be purchased for each exam attempt.

All candidates receive electronic certificates and personal listing on the CSWP directory when they pass.

Exam features hands-on challenges in many of these areas of the SOLIDWORKS Surfacing functionality:

Splines, 3D Curves, Boundary Surface, Filled Surface, Swept Surface, Planar Surface, Knit Surface, Trim Surface, Untrim Surface, Move Face, Extend-Surface, Fillets, and thicken.

CSWPA – Surfacing

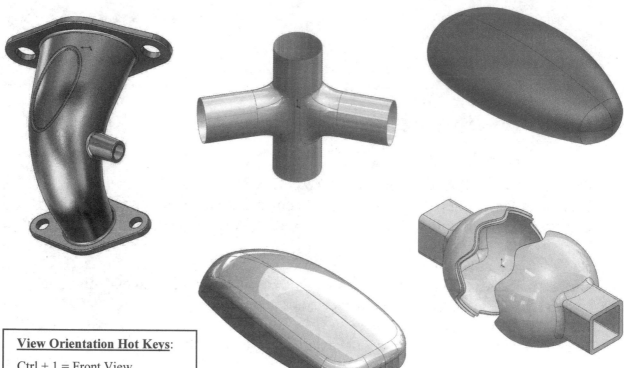

Dimensioning Standards: **ANSI**

Units: **INCHES** – 3 Decimals

Tools Needed:

 Extruded Surface   Planar Surface   Filled Surface

 Move/Copy   Knit Surface   Offset Surface

 Trim Surface   Cut with Surface   Mass Properties

CHALLENGE 1

1. Opening a part document:

- Select **File / Open**.

- Browse to the Training Folder and open the part document named **Inlet-Outlet Patches.sldprt**.

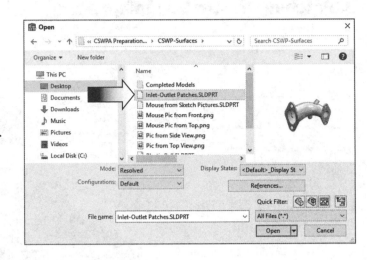

- This challenge examines your skills on creating 3D Curves, Boundary Surfaces, Filled Surfaces, Swept Surfaces, Planar Surfaces, Knit Surfaces, Trim Surfaces, Untrim Surfaces, Move Faces, Extend Surfaces, Fillets, and thicken.

2. Enabling the Surfaces tab:

- Right click the **Evaluate** tab and enable the **Surfaces** tab (arrow).

- There are two openings in this surface model.

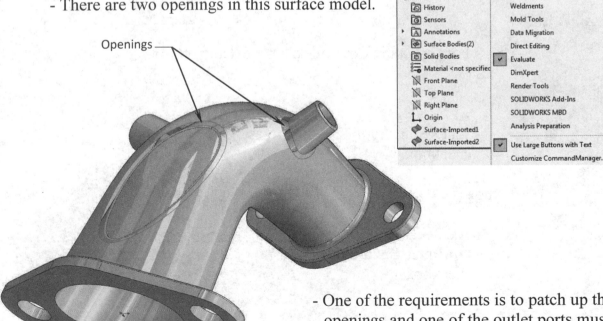

Openings

- One of the requirements is to patch up the openings and one of the outlet ports must be removed completely without leaving any trace of it.

3. Patching the first opening:

- Switch to the Surfaces tool tab.

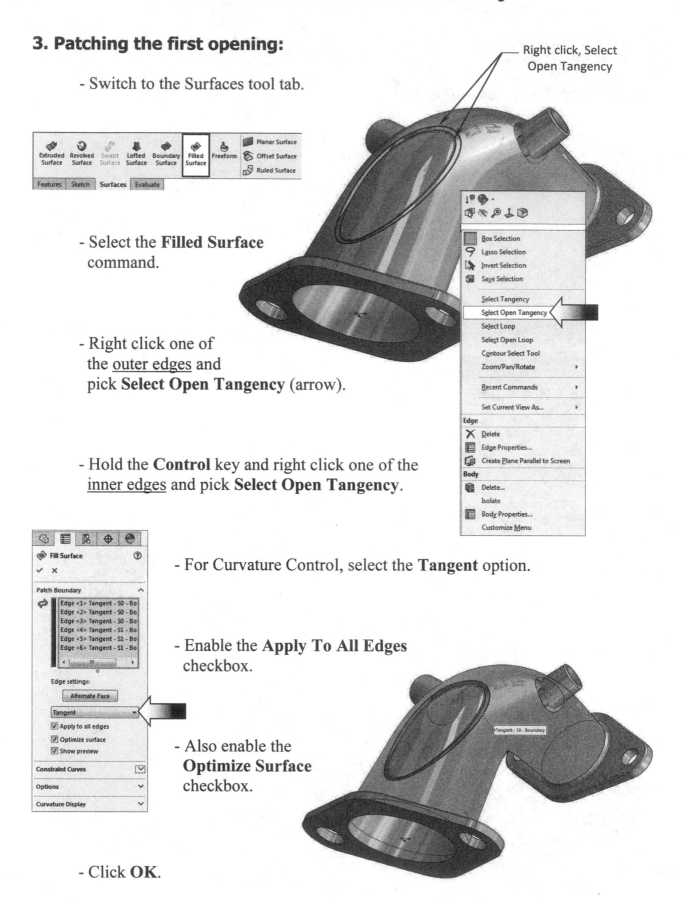

Right click, Select
Open Tangency

- Select the **Filled Surface** command.

- Right click one of the <u>outer edges</u> and pick **Select Open Tangency** (arrow).

- Hold the **Control** key and right click one of the <u>inner edges</u> and pick **Select Open Tangency**.

- For Curvature Control, select the **Tangent** option.

- Enable the **Apply To All Edges** checkbox.

- Also enable the **Optimize Surface** checkbox.

- Click **OK**.

4. Measuring the surface area:

- Switch to the **Evaluate** tab.

- Select the **Measure** command.

- Click the surface of the patch and enter the **Area** here:

_____ in^2.

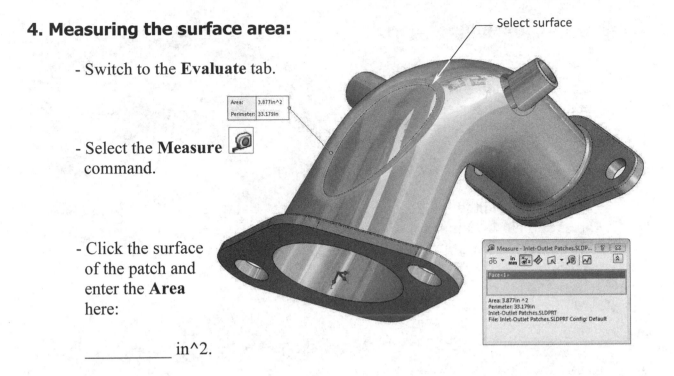

5. Deleting the outlet port:

- Select the **Delete Face** command from the **Surfaces** tool tab.

- Select the **Delete** option (arrow).

- Select all **5 faces** of the outlet port.

- Click **OK**.

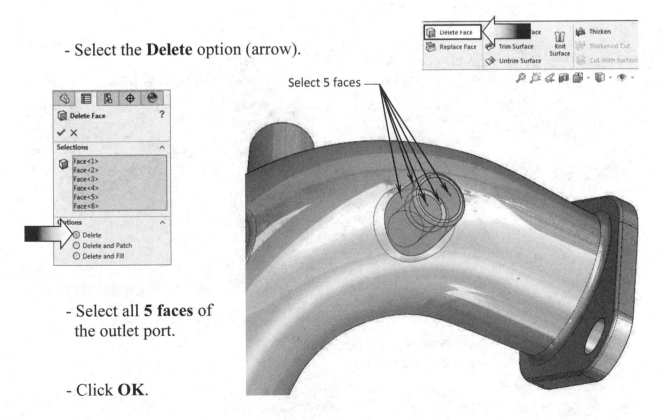

6. Filling the openings:

- One of the quick ways to fill
the openings left by the outlet
port is to delete the edges
of the openings.
SOLIDWORKS will
heal the openings
by extending the
surfaces around the
areas and close them
off without leaving any
traces.

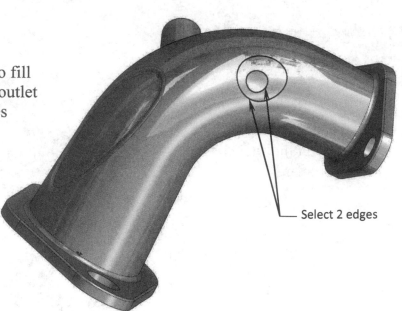

Select 2 edges

- Hold the **Control** key and
select the **2 edges** of the
2 openings, 1 on the inside
and 1 on the outside.

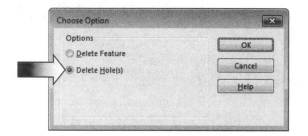

- Press the **Delete** key on the keyboard
and select the Delete Hole(s) option.

- Click **OK**.

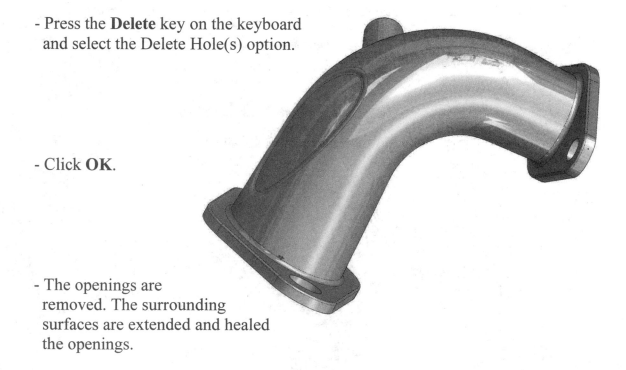

- The openings are
removed. The surrounding
surfaces are extended and healed
the openings.

7. Measuring the surface area:

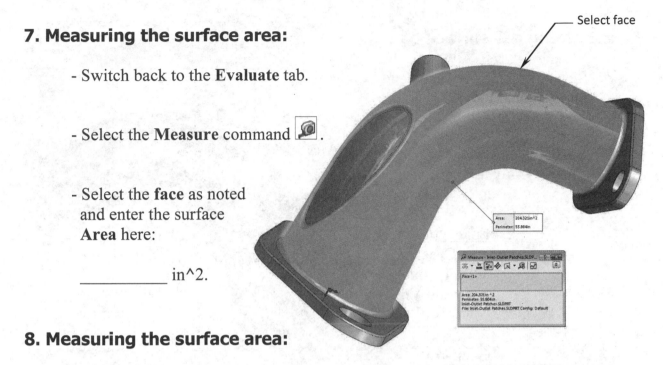

Select face

- Switch back to the **Evaluate** tab.

- Select the **Measure** command  .

- Select the **face** as noted and enter the surface **Area** here:

_____ in^2.

8. Measuring the surface area:

- Select the **Knit Surface** command from the **Surfaces** tool tab.

- Select the **3 surfaces** as indicated either from the graphics area or from the Surface Bodies folder.

- Enable the **Merge Entities** checkbox.

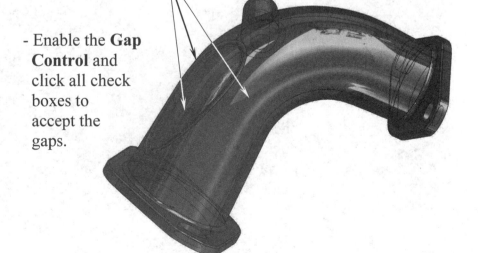

Select 3 surfaces

- Enable the **Gap Control** and click all check boxes to accept the gaps.

- Click **OK**.

9. Thickening the surface body:

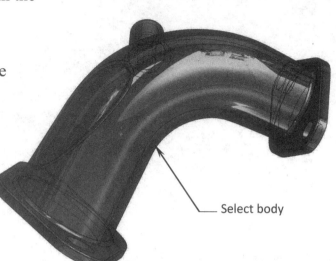

- Select the **Thicken** command from the **Surfaces** tool tab.

- For Thicken Parameters, select the **surface body** as noted.

- Enable the check box: **Create Solid From Enclosed Volume**.

- Click **OK**.

Select body

10. Creating a section view:

- Click the **Section View** command from the View Heads-Up toolbar.

- Use the **Top** plane as the Cutting plane.

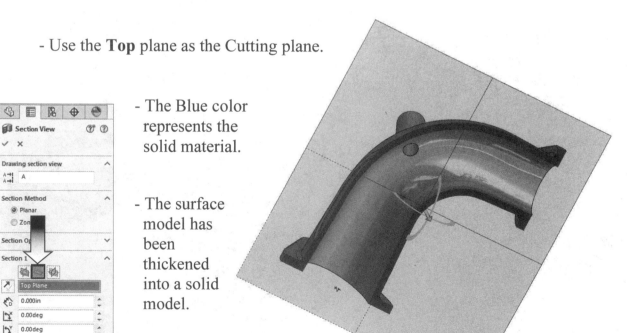

- The Blue color represents the solid material.

- The surface model has been thickened into a solid model.

- Push **Esc** to cancel the section view command.

11. Calculating the mass:

- Right click the Material option and select **Cast Alloy Steel** (arrow).

- Switch to the **Evaluate** tab.

- Click **Mass Properties** .

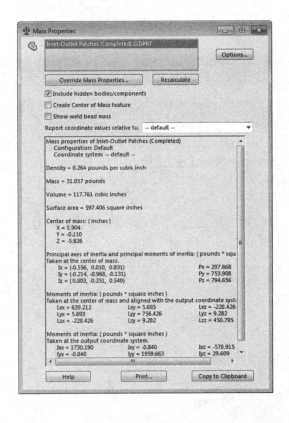

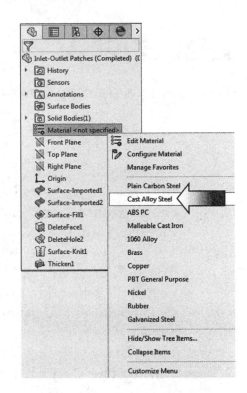

- Locate the mass of the part and enter it here:

_____ Pounds.

12. Saving your work:

- <u>Save</u> your work as **Inlet/Outlet (Completed).sldprt**.

CHALLENGE 2

1. Opening a part document:

- Select **File / Open**.

- Browse to the Training Folder and open the part document named **Mouse from Sketch Pictures.sldprt**.

- This challenge will examine your skills on working with sketch pictures and using various surface tools to create the body of a computer mouse.

2. Examining the sketch pictures:

- There are 2 sketch pictures that have been previously inserted into the Front and the Right planes.

- Expand the Sketch1 and the Sketch2 on the FeatureManager tree to see the two **Sketch Pictures** (arrow).

- The Sketch Picture1 represents the Front profile of the mouse body, and the Sketch Picture2 represents the Top.

3. Tracing the Top profile:

- The Top profile will be created out of 3 separate sketches. The first one is a spline traced along the upper curved edge of the Sketch Picture2.

- Select the **Top** plane and open a new sketch.

- Create a **2-Point Spline** along the edge as shown below.

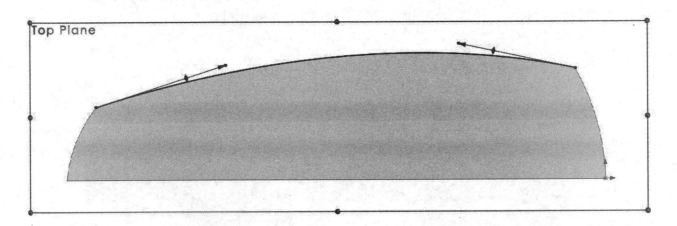

4. Extruding a surface:

- Switch to the **Surfaces** tool tab and click the **Extruded Surface** command.

- Use the default **Blind** type.

- Enter **1.000in** for depth.

- Extrude the surface upward.

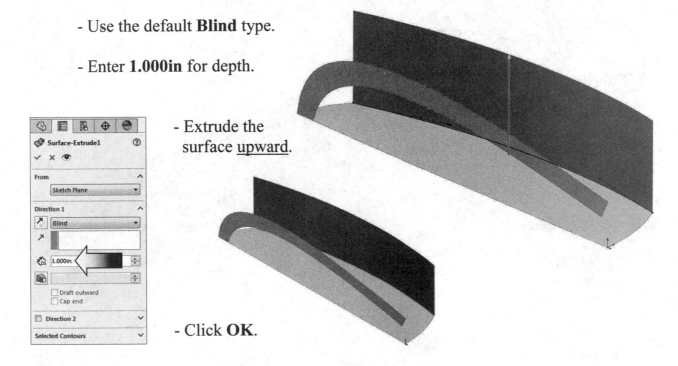

- Click **OK**.

5. Tracing the Front profile:

- Two sketches will be made from the Sketch Picture1 to create the Front profile.

- Select the **Front** plane and open a <u>new sketch</u>.

- Create a **2-Point Spline** along the <u>lower edge</u> as shown below.

- Add a **Horizontal Construction Line** (centerline) on the lower left side of the sketch picture. Also add the **.230in** dimension as noted.

- Drag the circular handles to adjust the spline.

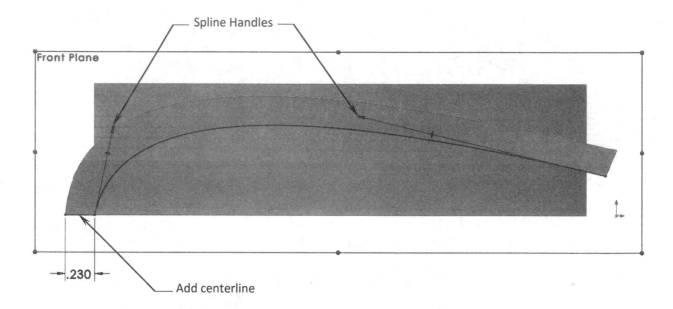

Spline Handles

Front Plane

.230

Add centerline

6. Extruding a surface:

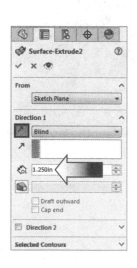

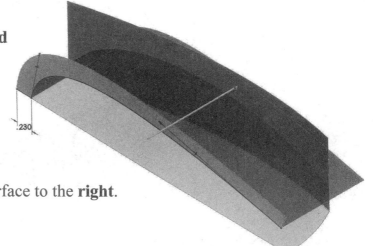

- Click **Extruded Surface**.

- Enter **1.250in** for depth.

- Extrude the surface to the **right**.

- Click **OK**.

7. Trimming the surfaces:

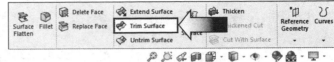

- Select the **Trim Surface**
  command from the Surfaces tool tab.

- Click the **Mutual** trim option (arrow).

- For Trim Selections, select the **2 Extruded Surfaces**.

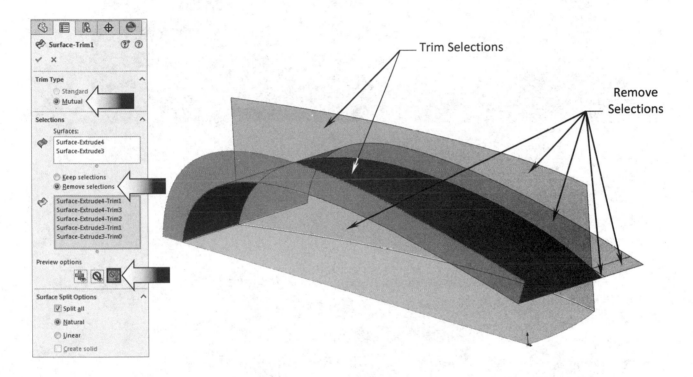

- For Remove Selections, select the **4 regions** of the 2 surfaces as noted.

- Select the option **Show Both Included and Excluded Surfaces** (arrow).

- Click **OK**.

8. Tracing the second sketch for the Front profile:

- Select the **Front** plane and open a new sketch.

- Create a **2-Point Spline** along the upper edge as shown below.

- Drag the circular handles to adjust the spline.

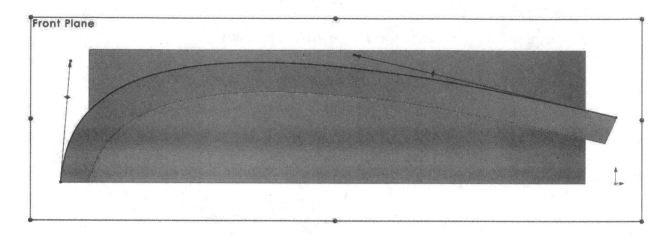

- **Exit** the Sketch.

9. Creating a new plane:

- Click the **Right** plane from the Feature tree. Hold the Control key and drag the right plane to the left side to make a copy from it.

- The Plane PropertyManager is displayed.

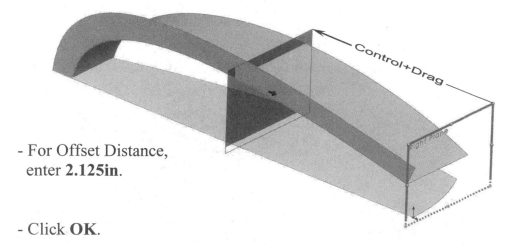

- For Offset Distance, enter **2.125in**.

- Click **OK**.

10. Creating the first boundary profile:

- Open a <u>new sketch</u> on the **Plane1**.

- Sketch a **2-Point Spline** and add a **Pierce** relation to <u>both ends</u> of the spline.

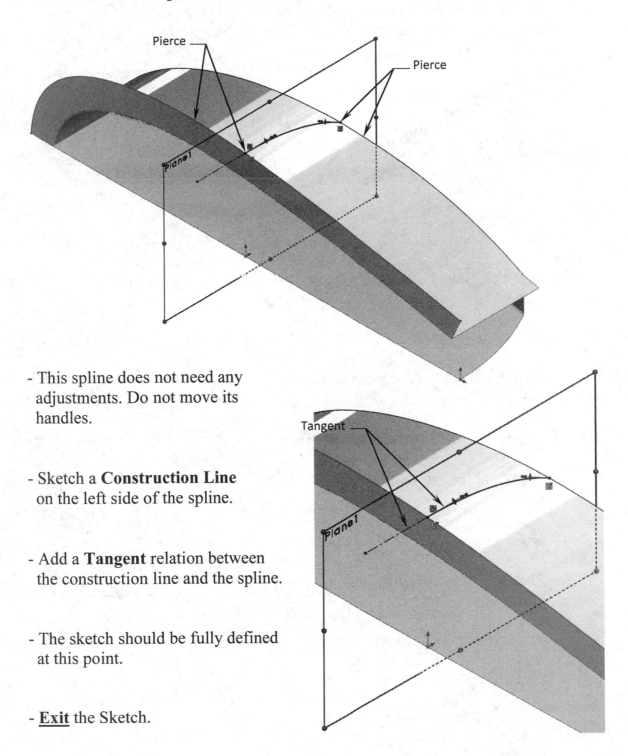

- This spline does not need any adjustments. Do not move its handles.

- Sketch a **Construction Line** on the left side of the spline.

- Add a **Tangent** relation between the construction line and the spline.

- The sketch should be fully defined at this point.

- <u>**Exit**</u> the Sketch.

11. Creating the second boundary profile:

- The second boundary profile is a 3D sketch.

- Select the **3D Sketch** command from the Sketch drop-down (arrow).

- Sketch a **2-Point Spline** as shown. Be sure to make both ends of the spline **Coincident** to the vertices of the 2 front and back surfaces.

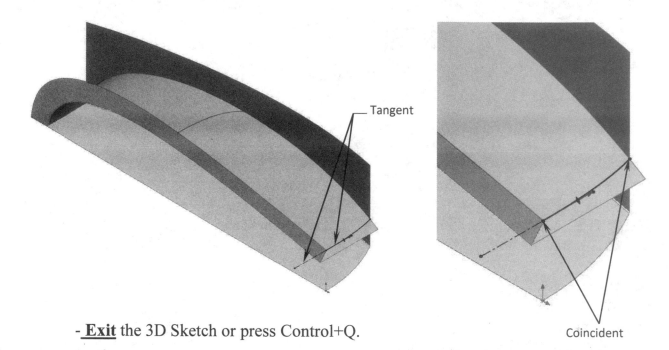

Tangent

Coincident

    - **Exit** the 3D Sketch or press Control+Q.

12. Creating the 3rd boundary profile:

- The 2nd boundary profile is a 2D sketch.

- Open a new sketch on the **Top** plane.

- Sketch a **2-Point Spline** on the left side of the Sketch Picture2.

- Ensure that both ends of the spline are Coincident to the vertices of the existing surfaces.

- **Exit** the Sketch.

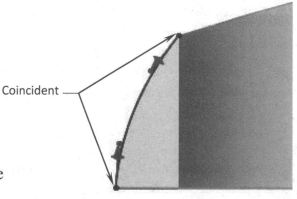

Coincident

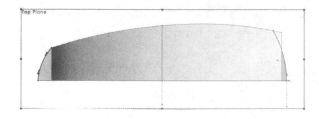

13. Hiding and suppressing the references:

- For clarity, we will hide some of the surfaces and suppress the 2 Sketch Pictures.

- Click the **Surface-Extrude1** and select **Hide** .

Hide this surface

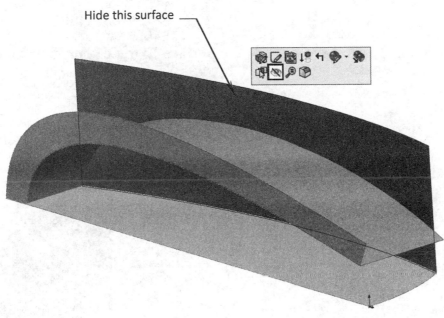

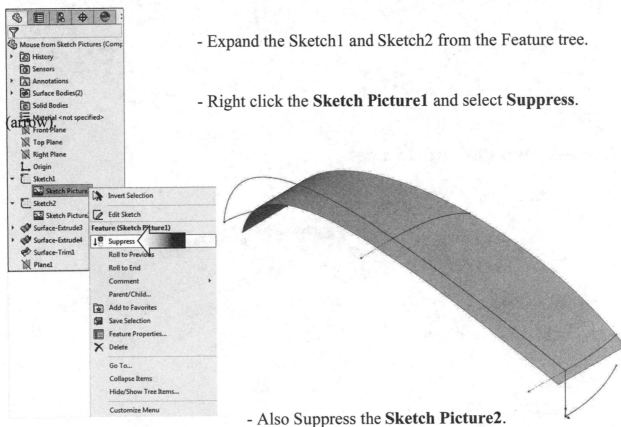

- Expand the Sketch1 and Sketch2 from the Feature tree.

- Right click the **Sketch Picture1** and select **Suppress**.

- Also Suppress the **Sketch Picture2**.

14. Creating the first Boundary Surface:

- Click the **Boundary Surface** command.

- For Direction 1, select the **3DSketch**1, **Sketch6** and **Sketch7** as noted.

- For Direction 2, select the **Edge** of the Surface-Extrude1 and the **Sketch5**.

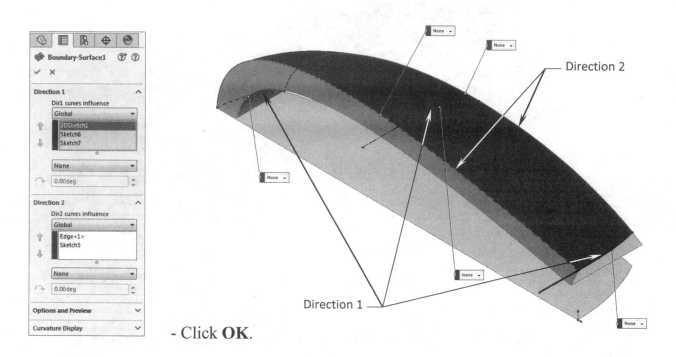

- Click **OK**.

15. Measuring the surface area:

- Switch to the **Evaluate** tab.

- Click **Measure** .

- Select the surface as indicated and enter the surface **Area** here:

_____ in^2.

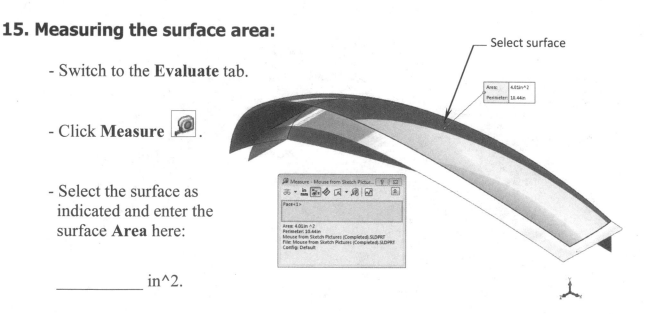

16. Creating another boundary profile:

- Select the **Right** plane and open a
new sketch.

- Sketch a **Line** from the origin to the
left corner of the boundary surface.

- The sketch should be fully defined
at this point.

- **Exit** the sketch.

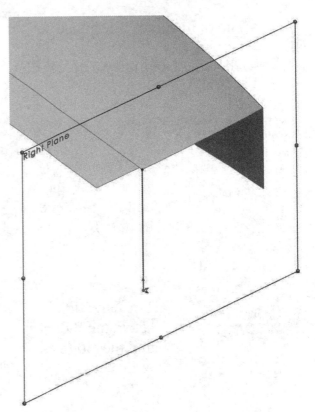

17. Sketching the next profile:

- Select the **Top** plane and open the
new sketch.

- Sketch a **2-Point Spline** from the
origin to the lower vertex of the
Surface Extrude1.

- The spline does
not need any
adjustments.
Do not move
the handles.

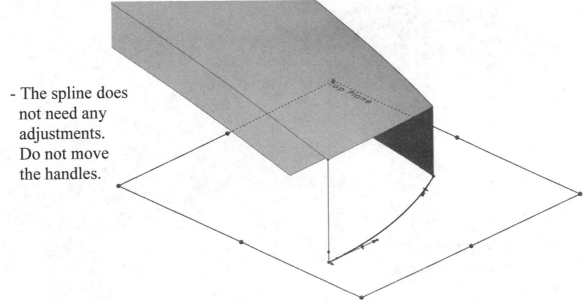

- **Exit** the sketch.

18. Creating the second boundary surface:

- The right end of the mouse body can be created with 4 simple lines/edges.

- Click the **Boundary Surface** command again.

- For Direction 1, select the **Edge** and the **Sketch8** as noted.

- For Direction 2, select the **Sketch9** and the **Edge** as indicated.

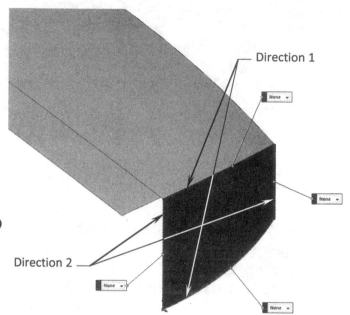

- Click **OK**.

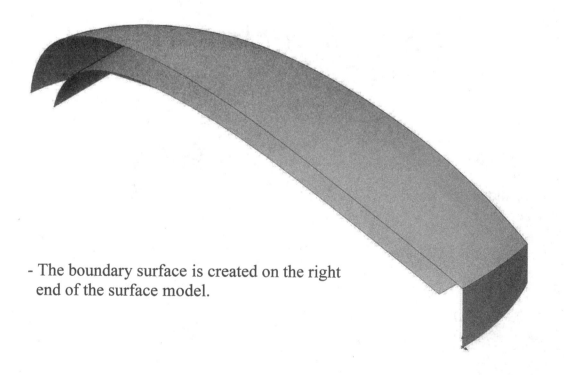

- The boundary surface is created on the right end of the surface model.

19. Creating a Surface Trim:

- Click the **Surface Trim** command 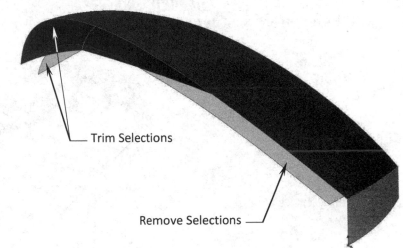.

- Select the **Mutual** option (arrow).

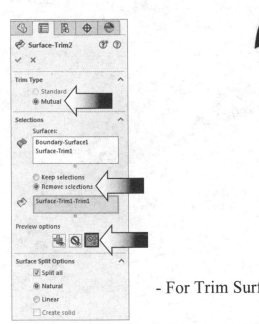

Trim Selections

Remove Selections

- For Trim Surfaces, select **upper** and **lower surfaces** as indicated.

- For remove Selection, select the lower surface as noted.

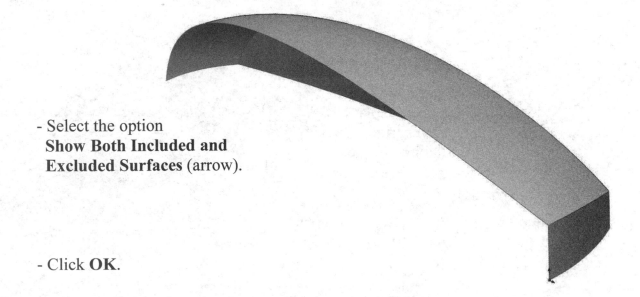

- Select the option
**Show Both Included and
Excluded Surfaces** (arrow).

- Click **OK**.

20. Creating a Curve Through Reference Points:

- A closed boundary is needed in order to create a planar surface. Either a line or a curve can be added to close off the bottom area.

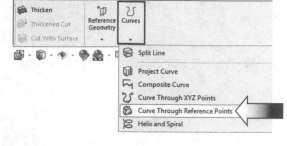

- Select the **Curve Through Reference Points** from the Curves drop-down.

- Select the **2 vertices** as indicated.

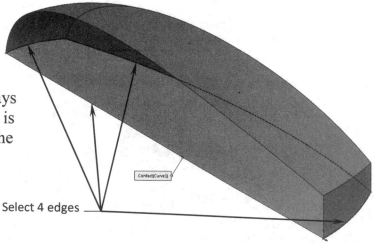

Select 2 vertices

- Click **OK**.

21. Creating a Filled Surface:

- Select the **Filled Surface** command.

- Select the **4 edges** as shown below.

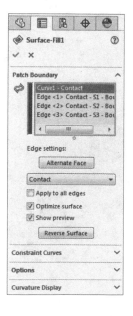

- The preview graphic displays a new surface is patching off the bottom area.

Contact(Curve1)

Select 4 edges

- Click **OK**.

22. Creating a Knit Surface:

- Click the **Knit Surface** command.

- Select the **4 surfaces** as noted.

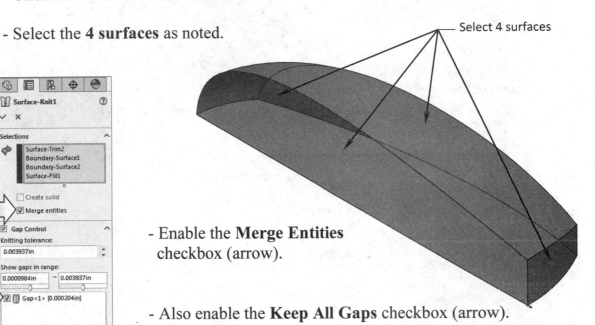

Select 4 surfaces

- Enable the **Merge Entities** checkbox (arrow).

- Also enable the **Keep All Gaps** checkbox (arrow).

- Click **OK**.

23. Mirroring the surface bodies:

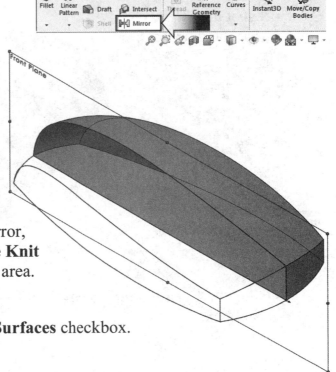

- Switch to the Features tool tab.

- Click the **Mirror** command.

- For Mirror Face/Plane, select the **Front** plane.

- For Bodies to Mirror, select the **Surface Knit** from the graphics area.

- Enable the **Knit Surfaces** checkbox.

- Click **OK**.

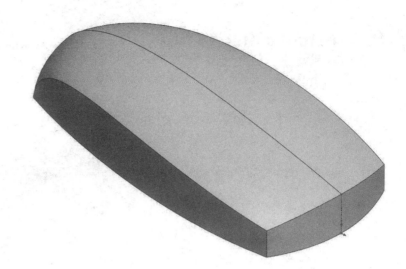

- The surface is mirrored and knitted into a single surface body.

24. Thickening the surface body.

- In order to calculate the mass for the model, it needs to be thickened into a solid Model.

- Click the **Thicken** command from the **Surfaces** tool tab.

- Select the **Surface** body from the graphics area.

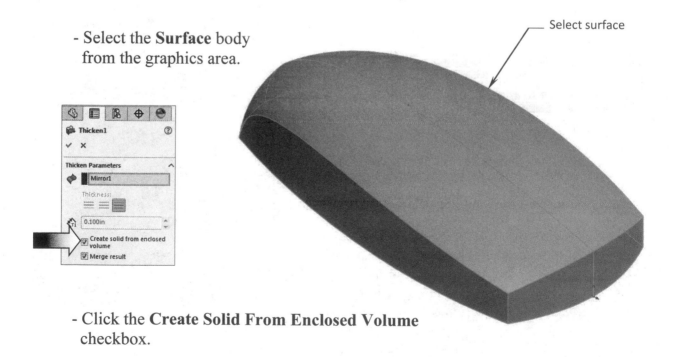

Select surface

- Click the **Create Solid From Enclosed Volume** checkbox.

- Click **OK**.

25. Creating a Section View:

- Create a section view to verify the solid model.

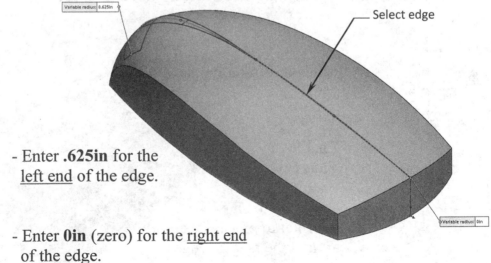

- Use the default **Front** plane as the Cutting Plane.

- The Blue color represents the solid body.

- Press **Esc** to exit the section view.

26. Adding a Variable Size fillet:

- Switch to the **Features** tool tab and select the **Fillet** command.

- Click the **Variable Size Fillet** option (arrow).

- Select the **edge** as noted.

Select edge

- Enter **.625in** for the left end of the edge.

- Enter **0in** (zero) for the right end of the edge.

- Click **OK**.

27. Adding a Constant Size Fillet:

- Select the **Fillet** command again.

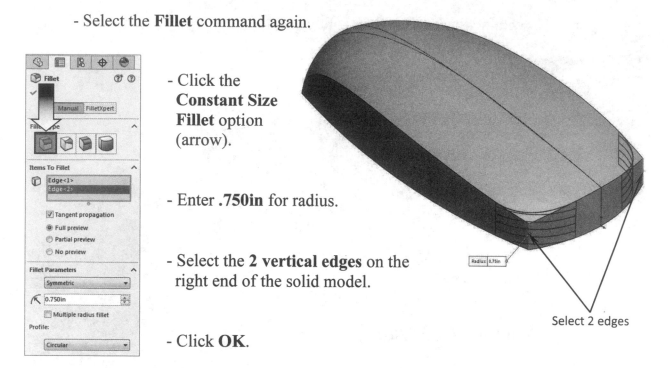

- Click the **Constant Size Fillet** option (arrow).

- Enter **.750in** for radius.

- Select the **2 vertical edges** on the right end of the solid model.

- Click **OK**.

Select 2 edges

Radius: 0.75in

28. Adding another Constant Size Fillet:

- Select the **Fillet** command once again.

- Keep the same **Constant Size Fillet option** and select the **edge** indicated.

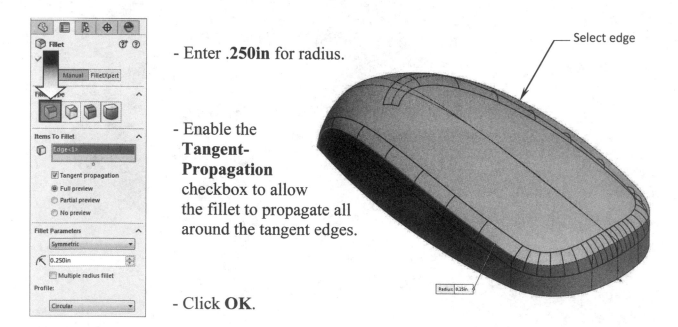

- Enter .**250in** for radius.

- Enable the **Tangent-Propagation** checkbox to allow the fillet to propagate all around the tangent edges.

- Click **OK**.

Select edge

Radius: 0.25in.

29. Adding a .032" Fillet:

- Select the **Fillet** command and select the **bottom edge** of the solid body.

- Enter **.032in** for radius.

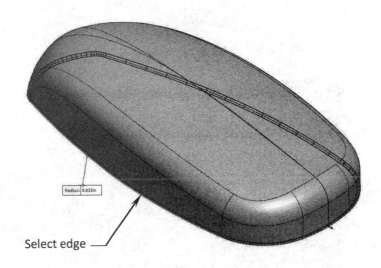

- The **Tangent Propagation** should still be enabled.

Radius: 0.032in

Select edge

- Click **OK**.

30. Assigning material:

- Right click the **Material** option and select **ABS PC** from the list.

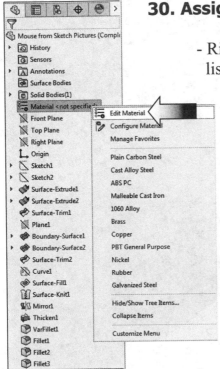

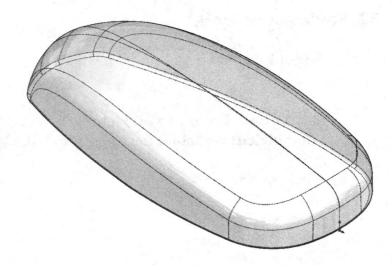

31. Calculating the mass:

- Switch to the **Evaluation** tab and click **Mass Properties** .

- Locate the mass of the part and enter it here:

_____ Pounds.

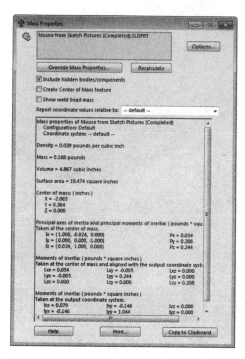

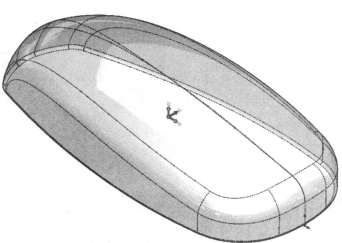

32. Saving your work:

- Select **File / Save As**.

-Enter **Mouse From Sketch Picture (Completed).sldprt** for the name of the file.

- Click **Save**.

- Close all documents.

CHALLENGE 3

1. Opening a part document:

- Select **File / Open**.

- Browse to the Training Folder and open the part document named **Plastic Ball.sldprt**.

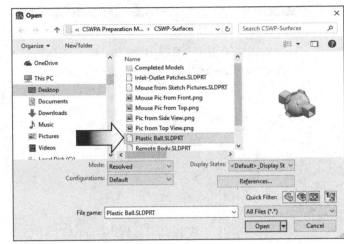

- This challenge examines your skills on the use of some of the surfacing tools such as Trim, Extend, Knit, Offset, Cut with Surface, etc.

2. Hiding the solid body:

- This model has 1 solid body and 2 surface bodies (arrow).

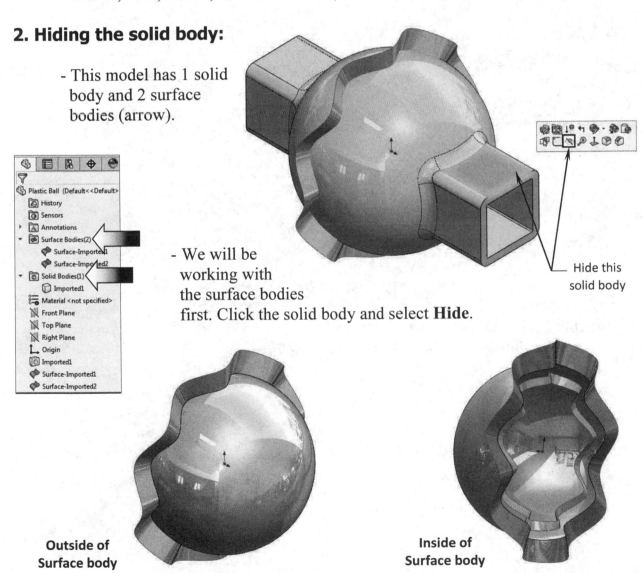

- We will be working with the surface bodies first. Click the solid body and select **Hide**.

Hide this solid body

Outside of Surface body

Inside of Surface body

3. Extending the Surface-Imported2:

- Switch to the **Surfaces** tool tab and click the **Surface Extend** command.

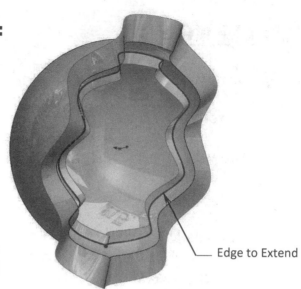

Edge to Extend

- For Edges/Faces to Extend, select the <u>outer edge</u> of the **Surface-Imported2** (the yellow surface).

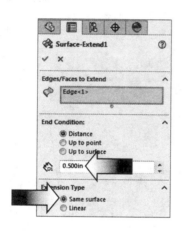

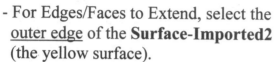

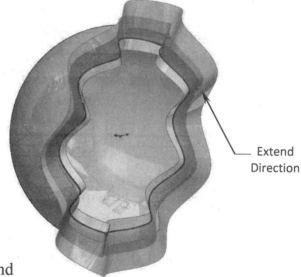

Extend Direction

- For End Condition, select **Distance** and enter **.500in.**, and extend the surface <u>outward</u>.

- Select the **Same Surface** option.

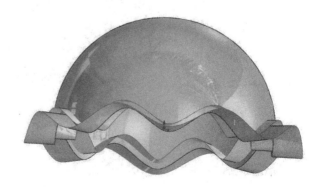

- Click **OK**.

4. Trimming the surfaces:

- Click the **Trim Surface** command.

- Select the **Mutual Trim** option (arrow).

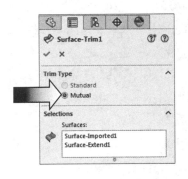

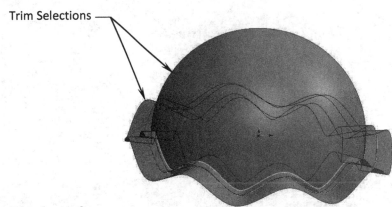

Trim Selections

- For Trim Selections, select both surfaces,
    the **Surface-Imported1** and the **Surface-Extend1**.

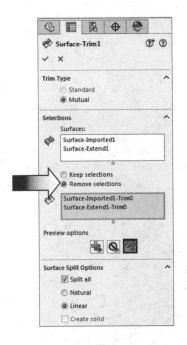

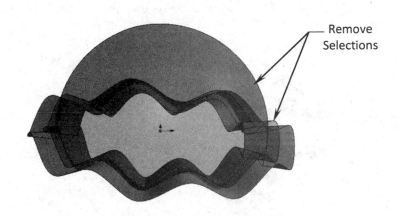

Remove
Selections

- For Remove Selections, select the **sphere** and the **extended
    portion** of the **Surface-Imported2**.

- Click the **Show Both Included and
    Excluded Surfaces** option (arrow).

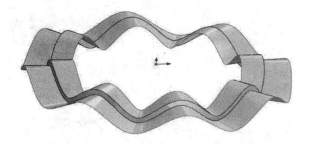

- Click **OK**.

5. Making a copy of the solid body:

- **<u>Show</u>** the solid body.

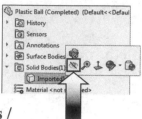

- Select **Insert / Features / Move/Copy**.

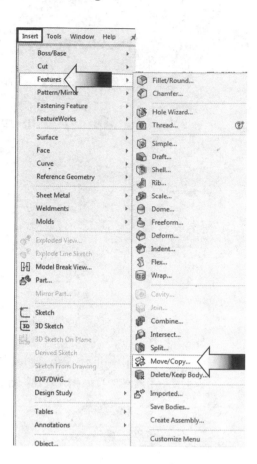

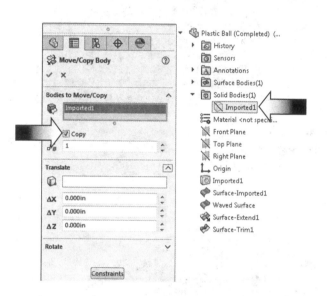

- Select the Solid Body (Imported1) either from the Feature tree or from the graphics area.

- Enable the **Copy** checkbox (arrow).

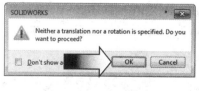

- Click **OK**.

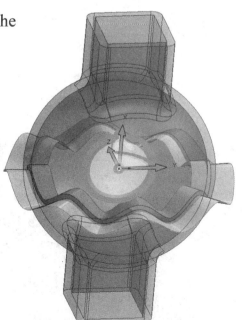

- Click **OK** again to proceed when prompted.

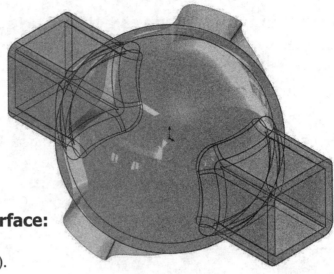

- The Body-Move/Copy1 is overlapping with the Imported1 body.

- The Surface-Trim is used to cut the two solid bodies into halves.

6. Creating the first Cut-With-Surface:

- Click **Cut With Surface** (arrow).

- For Surface Cut Parameter, select the **Surface-Trim1** either from the graphics area or from the Surface-Bodies folder.

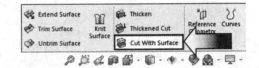

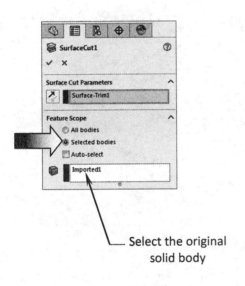

Select the original solid body

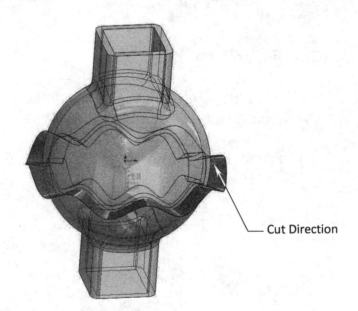

Cut Direction

- For Feature Scope, click the **Selected Body option**, then select the **Imported1** body.

- The Direction-Arrow points to the half that is going to be removed.

- Click **OK**.

7. Creating the second Cut-With-Surface:

- Click **Cut With Surface** (arrow).

- For Surface Cut Parameter, select the **Surface-Trim1** either from the graphics area or from the Surface-Bodies folder.

- For Feature Scope, click the Selected Body option, then select the **Body-Move/Copy1**.

- The Direction-Arrow points to the half that is going to be removed.

Cut Direction

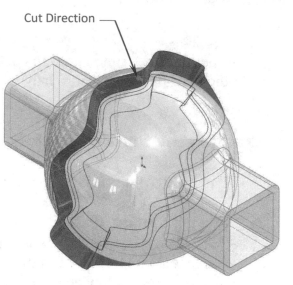

- Click **OK**.

8. Hiding a solid body:

- To see the result of the cuts, one of the bodies should be hidden.

- Expand the Surface Bodies folder and also the Solid Bodies folder (arrow).

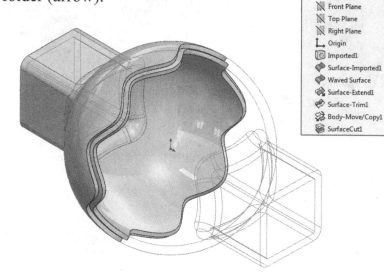

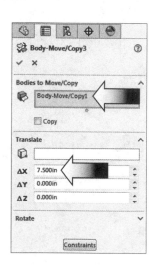

- **Hide** the **Surface-Trim1** and the **Body-Move/Copy1** (arrow).

9. Creating an Offset Surface:

- Click the **Offset Surface** command.

- Select the **face** of the Surface-Cut as noted.

- Enter **.060in** for Offset Distance.

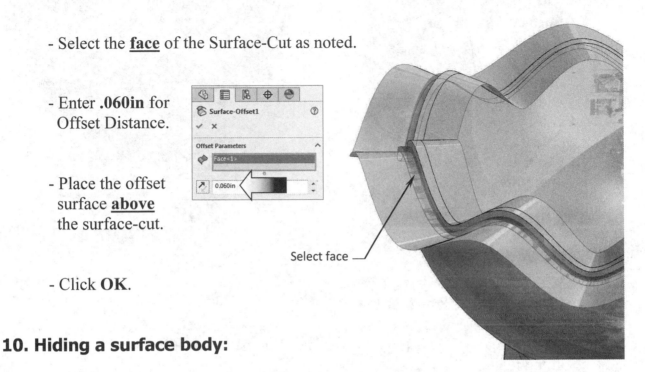

- Place the offset surface **above** the surface-cut.

Select face

- Click **OK**.

10. Hiding a surface body:

- We will need to hide the Trim-Surface so that we can work with the Offset Surface more easily.

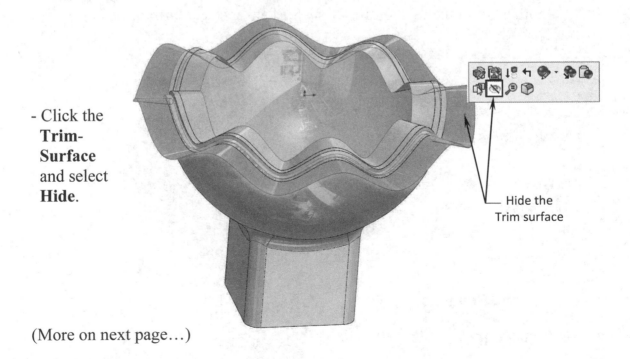

- Click the **Trim-Surface** and select **Hide**.

Hide the Trim surface

(More on next page...)

The new offset surface —

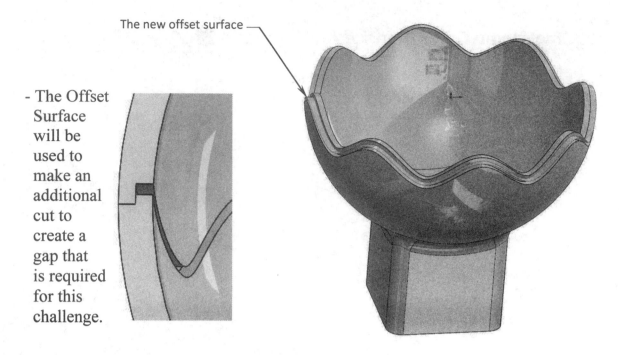

- The Offset Surface will be used to make an additional cut to create a gap that is required for this challenge.

11. Extending the Offset Surface:

- The Offset Surface is too narrow, it needs to be wider than the body that it is going to cut.

- Click the **Surface Extend** command.

- Either select the **face** of the Offset Surface <u>or</u> select **both** of its **edges**.

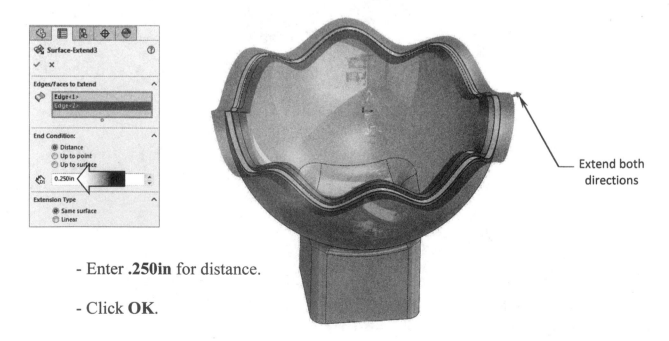

Extend both directions

- Enter **.250in** for distance.

- Click **OK**.

12. Creating the gap between the two solid bodies:

- Click the **Cut With Surface** command.

- For Surface Cut Parameter, select the new **Surface-Extend3**.

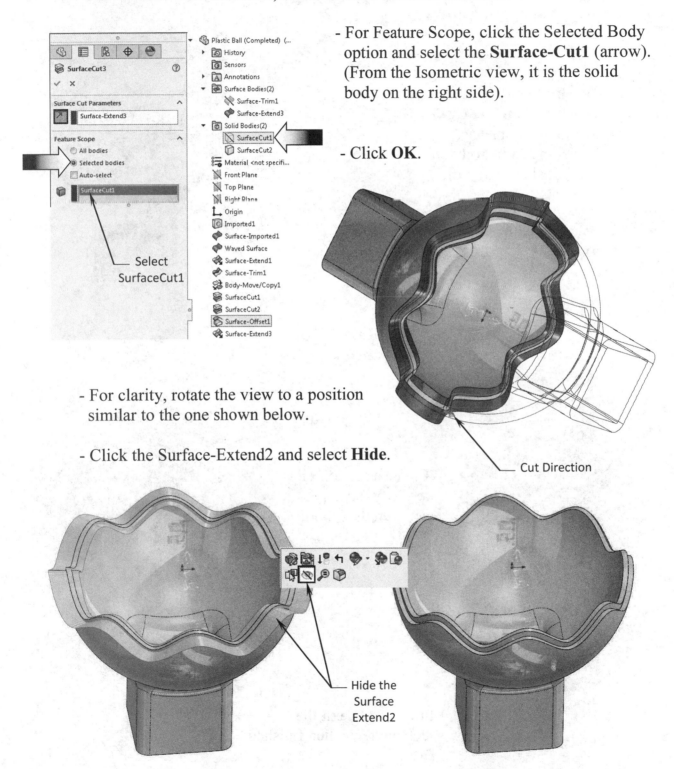

- For Feature Scope, click the Selected Body option and select the **Surface-Cut1** (arrow). (From the Isometric view, it is the solid body on the right side).

- Click **OK**.

Select SurfaceCut1

Cut Direction

- For clarity, rotate the view to a position similar to the one shown below.

- Click the Surface-Extend2 and select **Hide**.

Hide the Surface Extend2

13. Separating the solid bodies:

- Change to the **Features** tool tab.

- Select **Insert / Features / Move/Copy**.

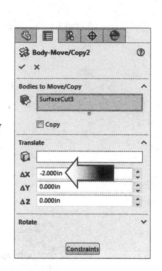

- For **Bodies to Move/Copy** select the **SurfaceCut3** (from the Isometric view, select the solid body on the left side to move).

- Enter **-2.000in** in the **Delta X** field.

- Click **OK**.

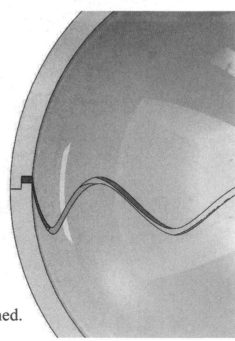

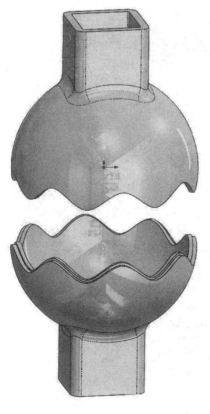

- Create a section view to verify the gap between the 2 bodies.

- Press <u>Esc</u> to exit the section view when finished.

14. Measuring the surface area:

- Switch to the **Evaluate** tab.

- Select the **Measure** command 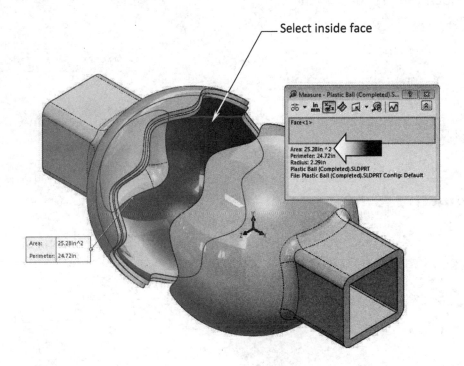.

- Select the **inside face** of the solid body on the left (**Body Move/Copy2**).

Select inside face

- Enter the surface Area here: _____ in^2.

15. Saving your work:

- Select **File / Save As**.

- Enter **Plastic Ball (completed).sldprt** for the file name.

- Click **Save**.

- Close all documents.

CHALLENGE 4

1. Opening a part document:

- Select **File / Open**.

- Browse to the Training Folder and open the part document named **Tube Patches.sldprt**.

- This challenge examines your skills on the use of some of the surfacing tools such as Trim Surface, Lofted Surface with Tangent Controls, and Filled Surface.

2. Examining the surface model:

- Beside the reference geometry that was saved in the Folder1, this surface model also comes with a sketch called Tube Centerline, a Plane3, and a Plane4.

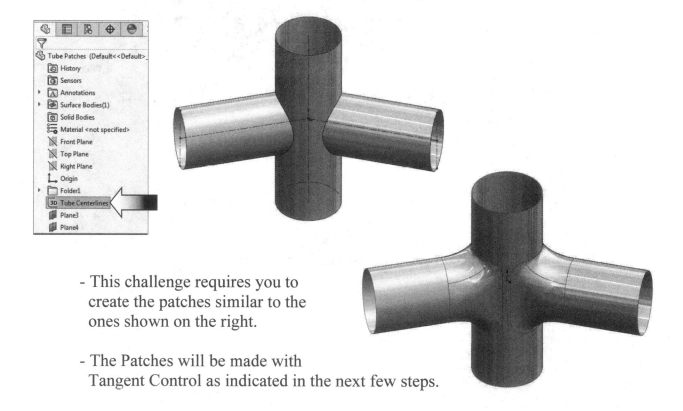

- This challenge requires you to create the patches similar to the ones shown on the right.

- The Patches will be made with Tangent Control as indicated in the next few steps.

3. Creating the trimming sketch:

- Select the **Plane3** and open
  a <u>new sketch</u>.

- Sketch the profile as shown.

- Add a vertical centerline from
  the Origin to use with the
  symmetric relations.

- Add the dimensions shown
  below to fully define the sketch.

(Note: There are 3 lines at the bottom).

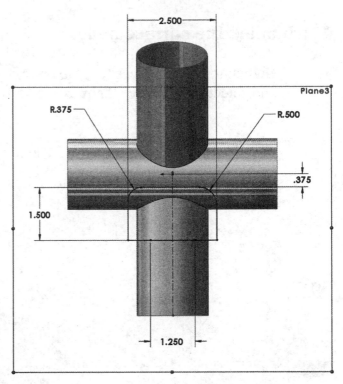

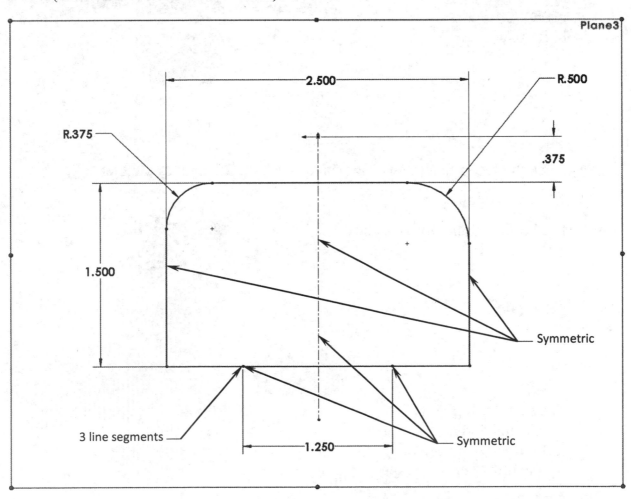

4. Trimming the surface body:

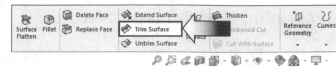

- Switch to the **Surfaces** tool tab
  and select the **Surface Trim** command.

- The **Standard** type should be the default when trimming with a sketch.

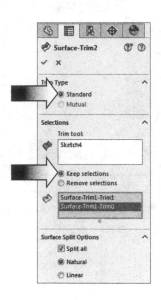

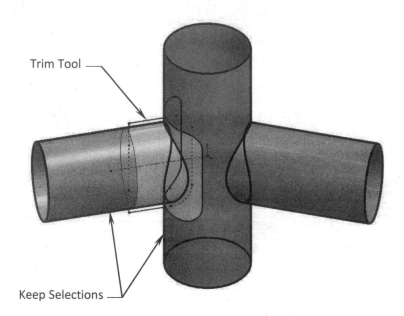

- Click the **Keep Selections** option and select the **2 surfaces** as noted.

- For Surface Split Options, click the
  Natural option.

- Click **OK**.

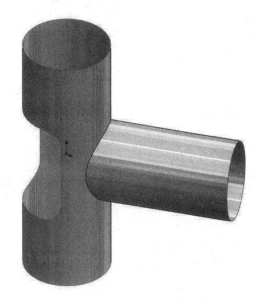

- The surface trim creates
  an opening and an additional
  surface body. We will fill the opening
  with a lofted surface in the next few steps.

5. Creating another trim surface:

- Select the **Plane4** and open a new sketch.

- Sketch the profile shown. Sketch **3 lines** on the **right side** of the profile as noted.

- Add a centerline from the Origin to use with the symmetric relations.

- Add dimensions to fully define the sketch.

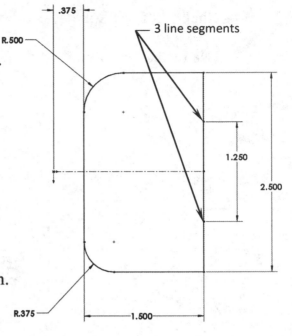

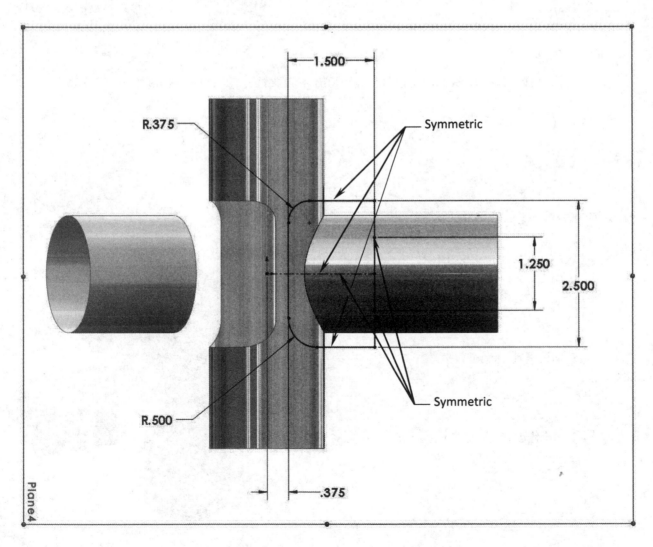

- Click the **Trim Surface** command .

- Use the default **Standard** trim type.

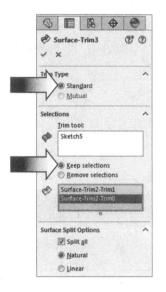

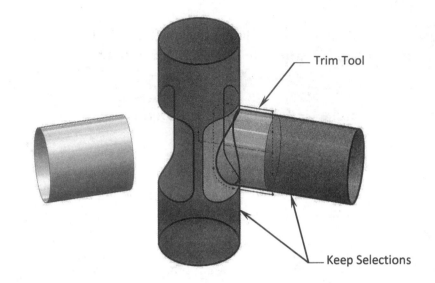

- Click the **Keep Selections** option and select the **2 surfaces** as noted.

- For Surface Split Options, select the **Natural** option.

- Click **OK**.

6. Measuring the surface area:

- Switch to the **Evaluate** tool tab.

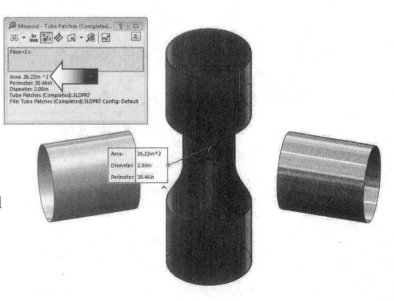

- Click **Measure**.

- Find the surface **Area** and enter it here:

_____ in^2.

7. Creating the first lofted surface:

- Either use Lofted Surface or Boundary Surface to create the patches as they produce much better quality and more accurate blends than other options.

- Click **Lofted Surface** 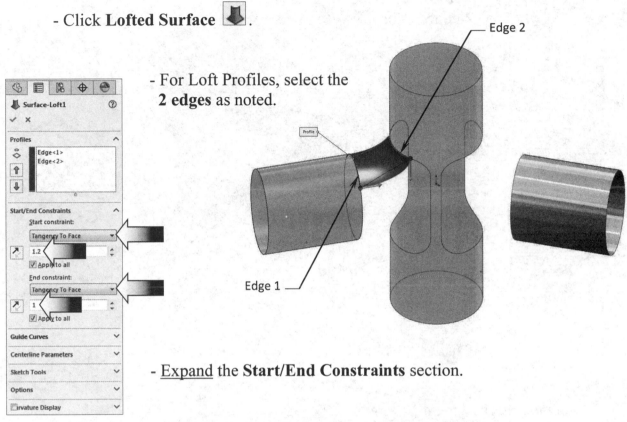.

- For Loft Profiles, select the **2 edges** as noted.

- Expand the **Start/End Constraints** section.

- Change the Start Constraint to **Tangency To Face**.

- Set the Start Tangent Length to **1.2**.

- Change the End Constraint to **Tangency To Face**.

- Leave the End Constraint value at **1**.

- Click **OK**.
  The first Lofted Surface is created.

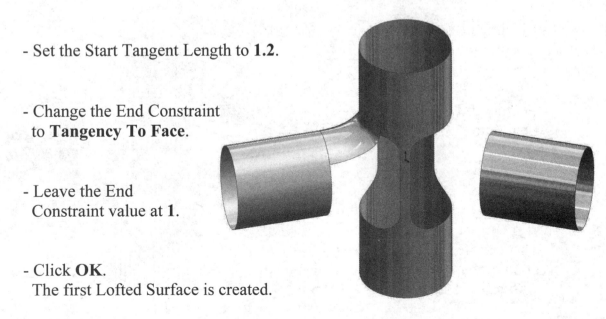

8. Creating the second lofted surface:

- Rotate the model to a similar orientation and click **Lofted Surface** .

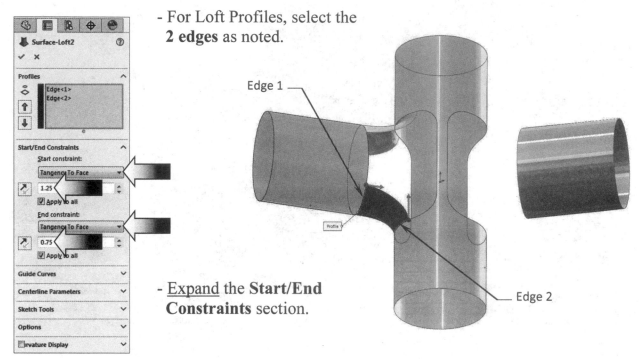

- For Loft Profiles, select the **2 edges** as noted.

Edge 1

Edge 2

- Expand the **Start/End Constraints** section.

- Change the Start Constraint to **Tangency To Face**.

- Set the Start Tangent Length to **1.25**.

- Change the End Constraint to **Tangency To Face**.

- Set the End Constraint Length to **.75**.

- Click **OK**.

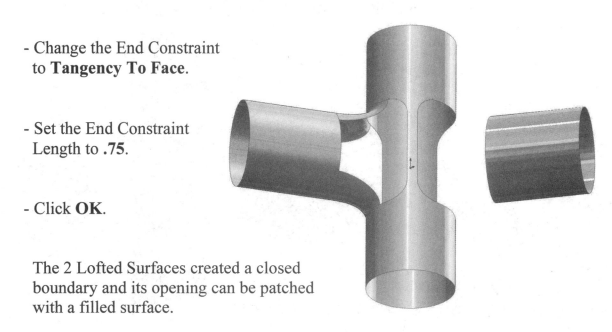

The 2 Lofted Surfaces created a closed boundary and its opening can be patched with a filled surface.

9. Creating the first filled surface:

- The Filled Surface command constructs a surface patch with any number of sides, within a boundary defined by existing model edges, sketches, or curves, including composite curves.

- Click **Filled Surface** .

- For Patch Boundary, select the **6 edges** of the opening as noted.

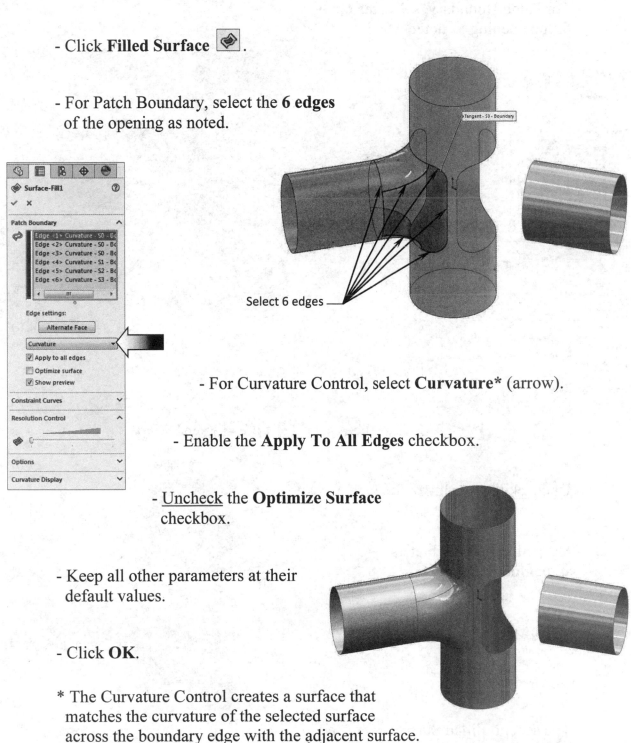

Select 6 edges

- For Curvature Control, select **Curvature*** (arrow).

- Enable the **Apply To All Edges** checkbox.

- Uncheck the **Optimize Surface** checkbox.

- Keep all other parameters at their default values.

- Click **OK**.

* The Curvature Control creates a surface that matches the curvature of the selected surface across the boundary edge with the adjacent surface.

10. Creating the second filled surface:

- Rotate the model around and work on the back side.

- For Patch Boundary, select the **6 edges** of the opening as noted.

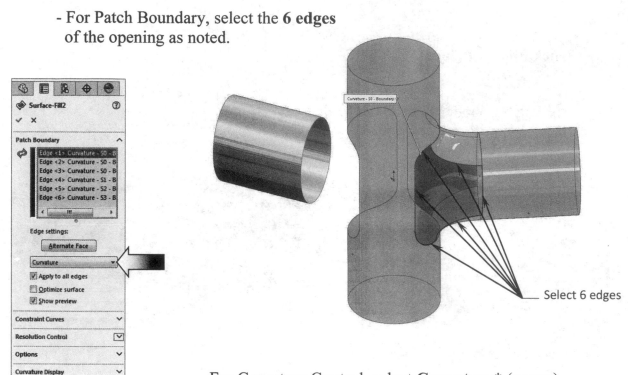

Select 6 edges

- For Curvature Control, select **Curvature*** (arrow).

- Enable the **Apply To All Edges** checkbox.

- <u>Uncheck</u> the **Optimize Surface** checkbox.

- Keep all other parameters at their default values.

- Click **OK**.

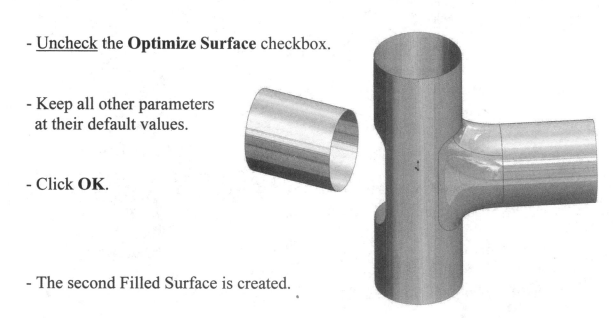

- The second Filled Surface is created.

11. Creating the third lofted surface:

- Click **Lofted Surface** ⬇.

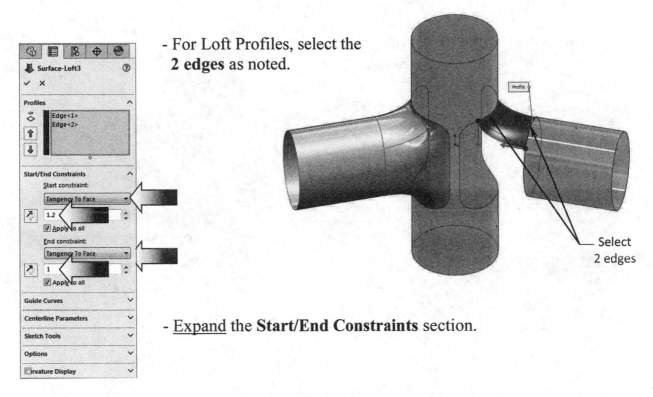

- For Loft Profiles, select the **2 edges** as noted.

Select 2 edges

- <u>Expand</u> the **Start/End Constraints** section.

- Change the Start Constraint to **Tangency To Face**.

- Set the Start Tangent Length to **1.2**.

- Change the End Constraint to **Tangency To Face**.

- Leave the End Constraint value at **1**.

- Click **OK**.

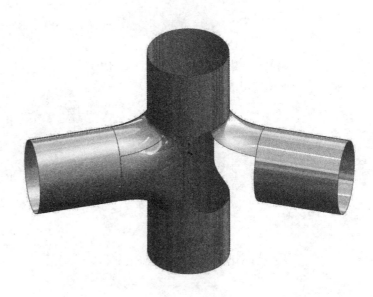

12. Creating the third lofted surface:

- Click **Lofted Surface** 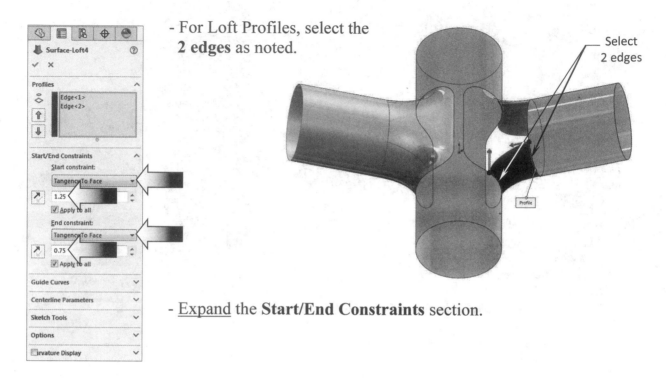.

- For Loft Profiles, select the **2 edges** as noted.

Select 2 edges

- Expand the **Start/End Constraints** section.

- Change the Start Constraint to **Tangency To Face**.

- Set the Start Tangent Length to **1.25**.

- Change the End Constraint to **Tangency To Face**.

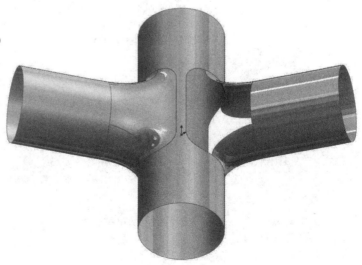

- Set the End Constraint Length to **.75**.

- Click **OK**.

13. Creating the third filled surface:

- Click **Lofted Surface** 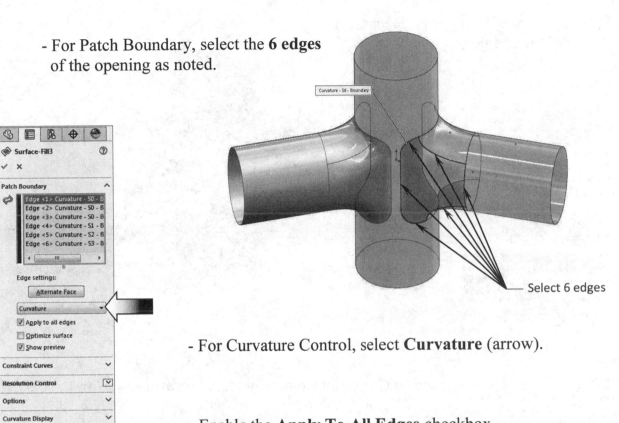.

- For Patch Boundary, select the **6 edges** of the opening as noted.

Select 6 edges

- For Curvature Control, select **Curvature** (arrow).

- Enable the **Apply To All Edges** checkbox.

- Uncheck the **Optimize Surface** checkbox.

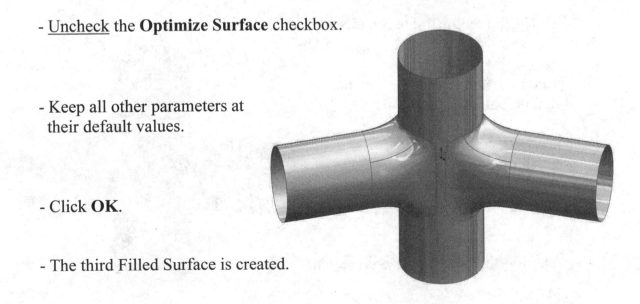

- Keep all other parameters at their default values.

- Click **OK**.

- The third Filled Surface is created.

14. Creating the fourth filled surface:

- Rotate the model to work on the back side.

- For Patch Boundary, select the **6 edges** of the opening as noted.

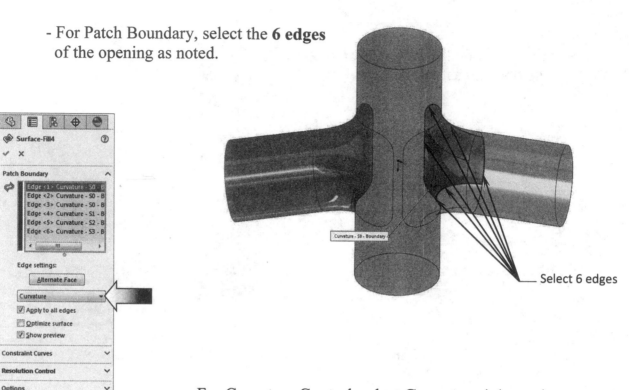

Select 6 edges

- For Curvature Control, select **Curvature*** (arrow).

- Enable the **Apply To All Edges** checkbox.

- <u>Uncheck</u> the **Optimize Surface** checkbox.

- Keep all other parameters at their default values.

- Click **OK**.

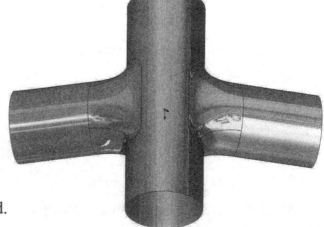

- The fourth Filled Surface is created.

15. Measuring the total surface area:

- Switch to the **Evaluate** tool tab.

- Click **Measure** 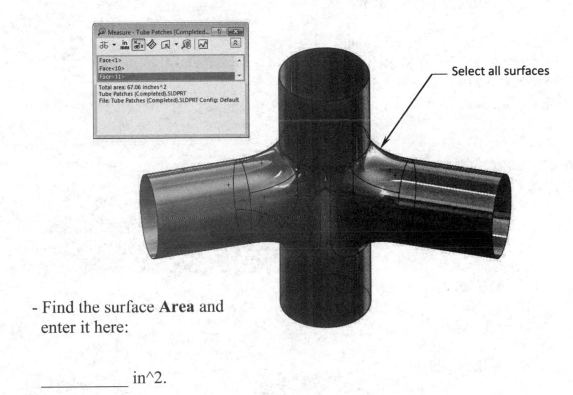 .

Select all surfaces

- Find the surface **Area** and
enter it here:

_____ in^2.

16. Saving your work:

- Select **File / Save As**.

- Enter **Tube Patches (Completed).sldprt** for the file name.

- Click **Save**.

- Close all documents.

CHALLENGE 5

1. Opening a part document:

- Select **File / Open**.

- Browse to the Training Folder and open the part document named **Remote Body.sldprt**.

- This challenge examines your skills on the use of Sketch-Picture, Splines, Lofted Surface, Knit Surface, and Thicken Surface.

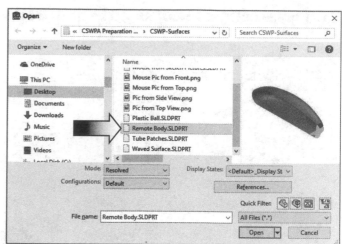

2. Tracing the first sketch picture:

- This surface model has two Sketch Pictures saved inside of Sketch1 and Sketch2.

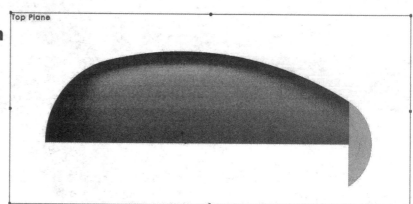

- **Show** both sketches.

- Select the **Top** plane and open a <u>new sketch</u>. We will trace the Sketch Picture1 first.

- Sketch a **2-Point Spline** as shown.

- Add a vertical centerline and two **Tangent** relations as indicated.

- Also add a **Horizontal** relation to the endpoints.
- <u>**Exit**</u> the sketch.

3. Tracing the second sketch picture:

- Select the **Front** plane and open a <u>new sketch</u>.

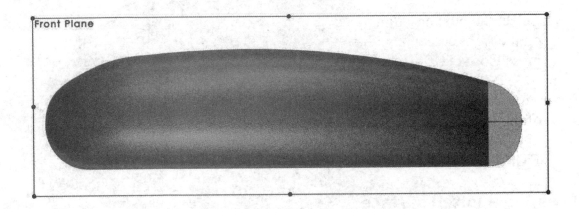

- Sketch a **3-Point-Spline** tracing along the outline of the Sketch Picture2.

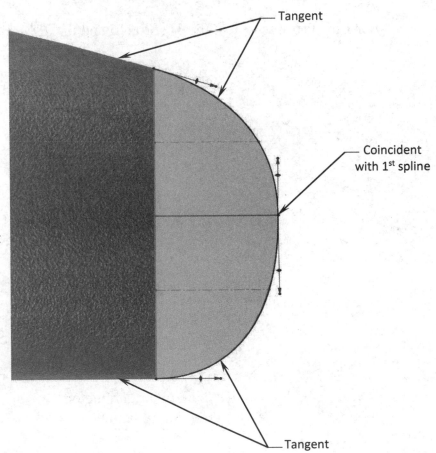

- Add two **Tangent** relations between the spline and the model edges.

- Add a **Coincident** relation between the middle spline point and the end point of the first spline.

- **Exit** the sketch.

4. Creating a 3D sketch:

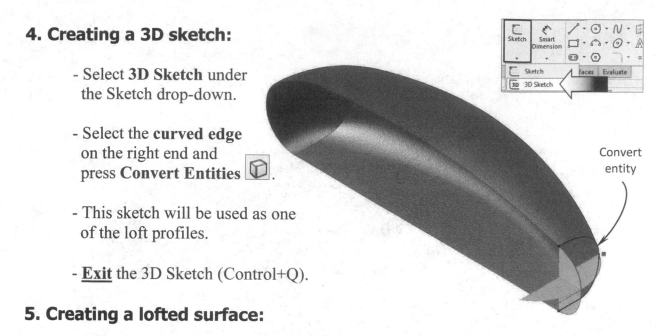

- Select **3D Sketch** under the Sketch drop-down.

- Select the **curved edge** on the right end and press **Convert Entities** 🔲.

- This sketch will be used as one of the loft profiles.

Convert entity

- <u>**Exit**</u> the 3D Sketch (Control+Q).

5. Creating a lofted surface:

- Switch to the **Surfaces** tool tab and select the **Lofted Surface** command ⬇.

- For Loft Profiles, select the **3D Sketch** and the **2ⁿᵈ Spline**.

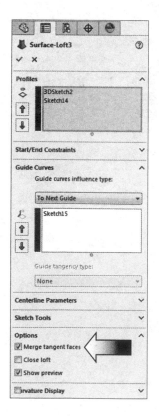

- For Guide Curve, select the **1ˢᵗ Spline**.

- Enable the **Merge Tangent Faces** check box.

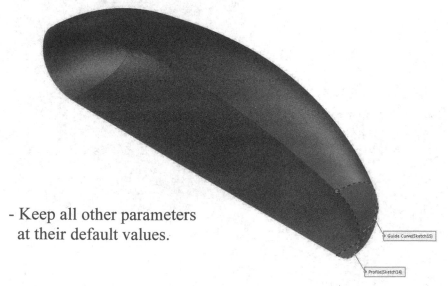

- Keep all other parameters at their default values.

- Click **OK**.

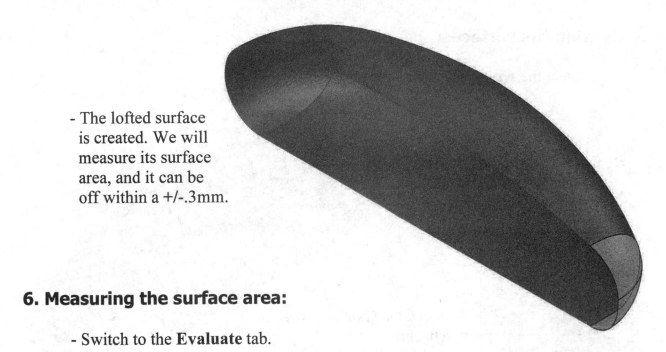

- The lofted surface
is created. We will
measure its surface
area, and it can be
off within a +/-.3mm.

6. Measuring the surface area:

- Switch to the **Evaluate** tab.

- Select the **Measure** command .

- Select **both halves** of the lofted surface.

- Find the surface
Area and enter
it here:

_____ in^2.

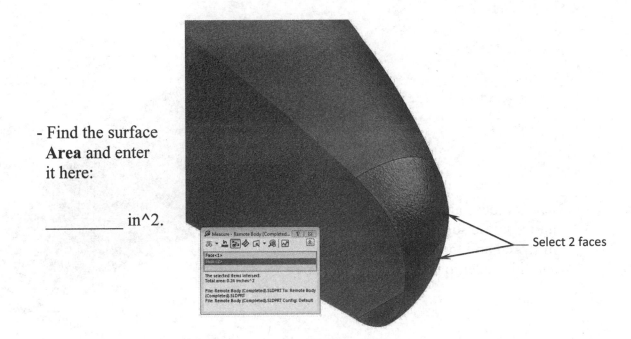

Select 2 faces

7. Knitting the surfaces:

- Select the **Knit Surface** command 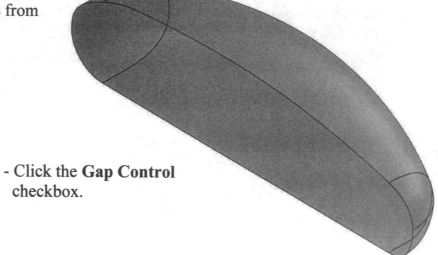 .

- Select **all surfaces** from
  the graphics area.

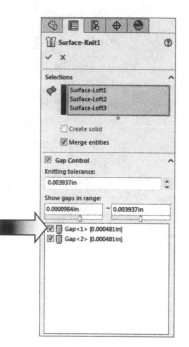

- Click the **Gap Control**
  checkbox.

- <u>Enable</u> the checkboxes to **keep all gaps** (arrow).

- Click **OK**.

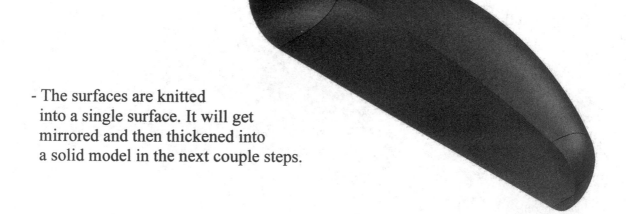

- The surfaces are knitted
  into a single surface. It will get
  mirrored and then thickened into
  a solid model in the next couple steps.

8. Mirroring the surface body:

- Switch to the **Feature** tool tab.

- Select the **Mirror** command.

- For Mirror Face/Plane, select the **Front** plane.

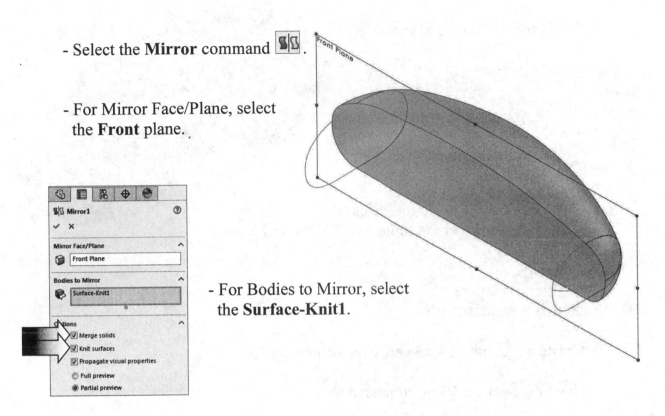

- For Bodies to Mirror, select the **Surface-Knit1**.

- Enable the **Merge Solid** and the **Knit Surfaces** check boxes (arrow)

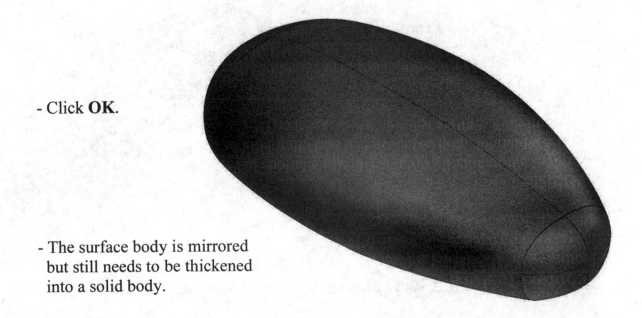

- Click **OK**.

- The surface body is mirrored but still needs to be thickened into a solid body.

9. Thickening the surface body:

- Switch back to the **Surfaces** tool tab.

- Select the **Thicken** command .

- For Thicken
Parameter, select
Mirror1 body from
the graphics area.

- Enable the checkbox
Create Solid From Enclosed Volume.

- Click **OK**.

10. Creating a section view:

- Create a section view to verify the solid body.

- Click the **Section View** command 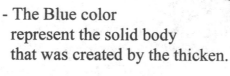.

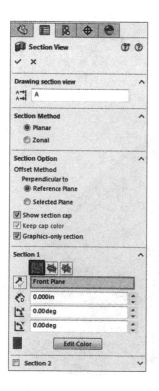

- Use the default
Front plane as
the Cutting plane.

- The Blue color
represent the solid body
that was created by the thicken.

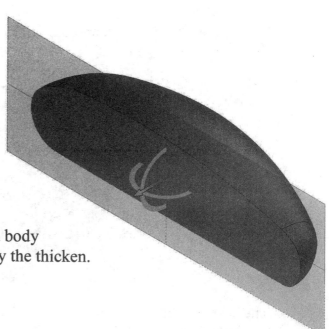

- Press **Esc** to exit the section view when finished.

11. Assigning material:

- Material needs to be assigned to the model in order to calculate the mass.

- Right click the Material option and select **Edit Material**.

- Select **ABS** from the **Plastics** folder.

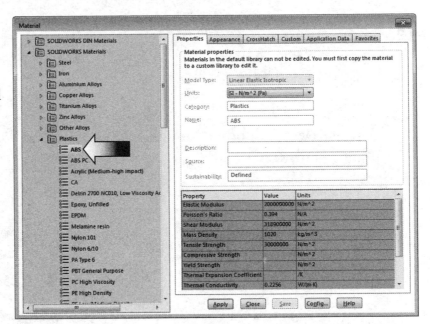

12. Calculating the final mass:

- Switch to the **Evaluate** tab and click **Mass Properties** .

- Locate the mass of the model and enter it here:

_____ Pounds.

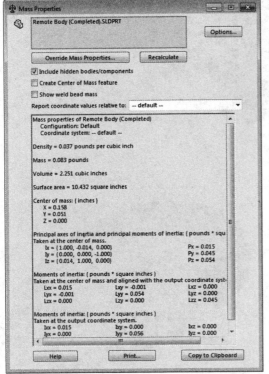

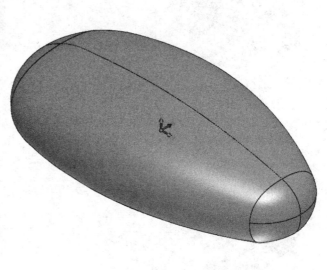

13. Saving your work

- Select **File / Save As**.

- Enter **Remote Body (Completed).sldprt** for the file name.

- Click **Save**.

- Close all documents.

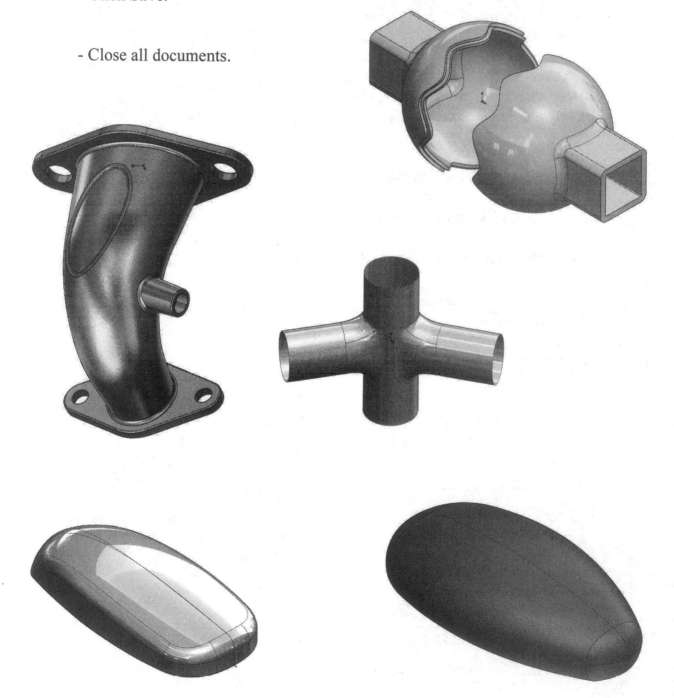

Glossary

Alloys:

An Alloy is a mixture of two or more metals (and sometimes a non-metal). The mixture is made by heating and melting the substances together.
Examples of alloys are Bronze (Copper and Tin), Brass (Copper and Zinc), and Steel (Iron and Carbon).

Gravity and Mass:

Gravity is the force that pulls everything on earth toward the ground and makes things feel heavy. Gravity makes all falling bodies accelerate at a constant 32ft. per second (9.8 m/s). In the earth's atmosphere, air resistance slows acceleration. Only on airless Moon would a feather and a metal block fall to the ground together.
The mass of an object is the amount of material it contains.
A body with greater mass has more inertia; it needs a greater force to accelerate.
Weight depends on the force of gravity, but mass does not.

When an object spins around another (for example, a satellite orbiting the earth) it is pushed outward. Two forces are at work here: Centrifugal (pushing outward) and Centripetal (pulling inward). If you whirl a ball around you on a string, you pull it inward (Centripetal force). The ball seems to pull outward (Centrifugal force) and if released will fly off in a straight line.

Heat:

Heat is a form of energy and can move from one substance to another in one of three ways: by Convection, by Radiation, and by Conduction.

- Convection takes place only in liquids like water (for example, water in a kettle) and gases (for example, air warmed by a heat source such as a fire or radiator). When liquid or gas is heated, it expands and becomes less dense. Warm air above the radiator rises and cool air moves in to take its place, creating a convection current.

- Radiation is movement of heat through the air. Heat forms match set molecules of air moving and rays of heat spread out around the heat source.

- Conduction occurs in solids such as metals. The handle of a metal spoon left in boiling liquid warms up as molecules at the heated end move faster and collide with their neighbors, setting them moving. The heat travels through the metal, which is a good conductor of heat.

Inertia:

A body with a large mass is harder to start and also to stop. A heavy truck traveling at 50mph needs more power brakes to stop its motion than a smaller car traveling at the same speed.
Inertia is the tendency of an object either to stay still or to move steadily in a straight line, unless another force (such as a brick wall stopping the vehicle) makes it behave differently.

Joules:

The Joules is the SI unit of work or energy.
One Joule of work is done when a force of one Newton moves over a distance of one meter. The Joule is named after the English scientist James Joule (1818-1889).

Materials:

- Stainless steel is an alloy of steel with chromium or nickel.

- Steel is made by the basic oxygen process. The raw material is about three parts melted iron and one part scrap steel. Blowing oxygen into the melted iron raises the temperature and gets rid of impurities.

- All plastics are chemical compounds called polymers.

- Glass is made by mixing and heating sand, limestone, and soda ash. When these ingredients melt they turn into glass, which is hardened when it cools. Glass is in fact not a solid but a "supercooled" liquid; it can be shaped by blowing, pressing, drawing, casting into molds, rolling, and floating across molten tin, to make large sheets.

- Ceramic objects, such as pottery and porcelain, electrical insulators, bricks, and roof tiles are all made from clay. The clay is shaped or molded when wet and soft, and heated in a kiln until it hardens.

Machine Tools:

Are powered tools used for shaping metal or other materials, by drilling holes, chiseling, grinding, pressing or cutting. Often the material (the work piece) is moved while the tool stays still (lathe), or vice versa, the work piece stays still while the tool moves (mill). Most common machine tools are Mill, Lathe, Saw, Broach, Punch press, Grind, Bore and Stamp break.

CNC

Computer Numerical Control is the automation of machine tools that are operated by precisely programmed commands encoded on a storage medium, as opposed to controlled manually via hand wheels or levers, or mechanically automated via cams alone. Most CNC today is computer numerical control in which computers play an integral part of the control.

3D Printing

All methods work by working in layers, adding material, etc. different to other techniques, which are subtractive. Support is needed because almost all methods could support multi material printing, but it is currently only available in certain top tier machines.

A method of turning digital shapes into physical objects. Due to its nature, it allows us to accurately control the shape of the product. The drawback is size restraints and materials are often not durable.

While FDM doesn't seem like the best method for instrument manufacturing, it is one of the cheapest and most universally available methods.

EDM
Electric Discharge Machining.

FDM
Fused Deposition Modeling.

SLA
Stereo Lithography.

SLS
Selective Laser Sintering.

SLM
Selective Laser Melting.

J-P
Jetted Photopolymer (or Polyjet)

Newton's Law:

1. Every object remains stopped or goes on moving at a steady rate in a straight line unless acted upon by another force. This is the inertia principle.
2. The amount of force needed to make an object change its speed depends on the mass of the object and the amount of the acceleration or deceleration required.
3. To every action there is an equal and opposite reaction. When a body is pushed on way by a force, another force pushes back with equal strength.

Polymers:

A polymer is made of one or more large molecules formed from thousands of smaller molecules. Rubber and Wood are natural polymers. Plastics are synthetic (artificially made) polymers.

Speed and Velocity:

- Speed is the rate at which a moving object changes position (how far it moves in a fixed time).
- Velocity is speed in a particular direction.
- If either speed or direction is changed, velocity also changes.

Absorbed

A feature, sketch, or annotation that is contained in another item (usually a feature) in the FeatureManager design tree. Examples are the profile sketch and profile path in a base-sweep, or a cosmetic thread annotation in a hole.

Align

Tools that assist in lining up annotations and dimensions (left, right, top, bottom, and so on). For aligning parts in an assembly.

Alternate position view

A drawing view in which one or more views are superimposed in phantom lines on the original view. Alternate position views are often used to show range of motion of an assembly.

Anchor point

The end of a leader that attaches to the note, block, or other annotation. Sheet formats contain anchor points for a bill of materials, a hole table, a revision table, and a weldment cut list.

Annotation

A text note or a symbol that adds specific design intent to a part, assembly, or drawing. Specific types of annotations include note, hole callout, surface finish symbol, datum feature symbol, datum target, geometric tolerance symbol, weld symbol, balloon, and stacked balloon. Annotations that apply only to drawings include center mark, annotation centerline, area hatch, and block.

Appearance callouts

Callouts that display the colors and textures of the face, feature, body, and part under the entity selected and are a shortcut to editing colors and textures.

Area hatch

A crosshatch pattern or fill applied to a selected face or to a closed sketch in a drawing.

Assembly

A document in which parts, features, and other assemblies (sub-assemblies) are mated together. The parts and sub-assemblies exist in documents separate from the assembly. For example, in an assembly, a piston can be mated to other parts, such as a connecting rod or cylinder. This new assembly can then be used as a sub-assembly in an assembly of an engine. The extension for a SOLIDWORKS assembly file name is .SLDASM.

Attachment point

The end of a leader that attaches to the model (to an edge, vertex, or face, for example) or to a drawing sheet.

Axis

A straight line that can be used to create model geometry, features, or patterns. An axis can be made in a number of different ways, including using the intersection of two planes.

Balloon

Labels parts in an assembly, typically including item numbers and quantity. In drawings, the item numbers are related to rows in a bill of materials.

Base

The first solid feature of a part.

Baseline dimensions

Sets of dimensions measured from the same edge or vertex in a drawing.

Bend

A feature in a sheet metal part. A bend generated from a filleted corner, cylindrical face, or conical face is a round bend; a bend generated from sketched straight lines is a sharp bend.

Bill of materials

A table inserted into a drawing to keep a record of the parts used in an assembly.

Block

A user-defined annotation that you can use in parts, assemblies, and drawings. A block can contain text, sketch entities (except points), and area hatch, and it can be saved in a file for later use as, for example, a custom callout or a company logo.

Bottom-up assembly

An assembly modeling technique where you create parts and then insert them into an assembly.

Broken-out section

A drawing view that exposes inner details of a drawing view by removing material from a closed profile, usually a spline.

Cavity

The mold half that holds the cavity feature of the design part.

Center mark

A cross that marks the center of a circle or arc.

Centerline

A centerline marks, in phantom font, an axis of symmetry in a sketch or drawing.

Chamfer

Bevels a selected edge or vertex. You can apply chamfers to both sketches and features.

Child

A dependent feature related to a previously-built feature. For example, a chamfer on the edge of a hole is a child of the parent hole.

Click-release

As you sketch, if you click and then release the pointer, you are in click-release mode. Move the pointer and click again to define the next point in the sketch sequence.

Click-drag

As you sketch, if you click and drag the pointer, you are in click-drag mode. When you release the pointer, the sketch entity is complete.

Closed profile

Also called a closed contour, it is a sketch or sketch entity with no exposed endpoints; for example, a circle or polygon.

Collapse

The opposite of explode. The collapse action returns an exploded assembly's parts to their normal positions.

Collision Detection

An assembly function that detects collisions between components when components move or rotate. A collision occurs when an entity on one component coincides with any entity on another component.

Component

Any part or sub-assembly within an assembly.

Configuration

A variation of a part or assembly within a single document. Variations can include different dimensions, features, and properties. For example, a single part such as a bolt can contain different configurations that vary the diameter and length.

ConfigurationManager

Located on the left side of the SOLIDWORKS window, it is a means to create, select, and view the configurations of parts and assemblies.

Constraint

The relations between sketch entities, or between sketch entities and planes, axes, edges, or vertices.

Construction geometry

The characteristic of a sketch entity that the entity is used in creating other geometry but is not itself used in creating features.

Coordinate system

A system of planes used to assign Cartesian coordinates to features, parts, and assemblies. Part and assembly documents contain default coordinate systems; other coordinate systems can be defined with reference geometry. Coordinate systems can be used with measurement tools and for exporting documents to other file formats.

Cosmetic thread

An annotation that represents threads.

Crosshatch

A pattern (or fill) applied to drawing views such as section views and broken-out sections.

Curvature

Curvature is equal to the inverse of the radius of the curve. The curvature can be displayed in different colors according to the local radius (usually of a surface).

Cut

A feature that removes material from a part by such actions as extrude, revolve, loft, sweep, thicken, cavity, and so on.

Dangling

A dimension, relation, or drawing section view that is unresolved. For example, if a piece of geometry is dimensioned, and that geometry is later deleted, the dimension becomes dangling.

Degrees of freedom

Geometry that is not defined by dimensions or relations is free to move. In 2D sketches, there are three degrees of freedom: movement along the X and Y axes, and rotation about the Z axis (the axis normal to the sketch plane). In 3D sketches and in assemblies, there are six degrees of freedom: movement along the X, Y, and Z axes, and rotation about the X, Y, and Z axes.

Derived part

A derived part is a new base, mirror, or component part created directly from an existing part and linked to the original part such that changes to the original part are reflected in the derived part.

Derived sketch

A copy of a sketch, in either the same part or the same assembly that is connected to the original sketch. Changes in the original sketch are reflected in the derived sketch.

Design Library

Located in the Task Pane, the Design Library provides a central location for reusable elements such as parts, assemblies, and so on.

Design table

An Excel spreadsheet that is used to create multiple configurations in a part or assembly document.

Detached drawing

A drawing format that allows opening and working in a drawing without loading the corresponding models into memory. The models are loaded on an as-needed basis.

Detail view

A portion of a larger view, usually at a larger scale than the original view.

Dimension line

A linear dimension line references the dimension text to extension lines indicating the entity being measured. An angular dimension line references the dimension text directly to the measured object.

DimXpertManager

Located on the left side of the SOLIDWORKS window, it is a means to manage dimensions and tolerances created using DimXpert for parts according to the requirements of the ASME Y.14.41-2003 standard.

DisplayManager

The DisplayManager lists the appearances, decals, lights, scene, and cameras applied to the current model. From the DisplayManager, you can view applied content, and add, edit, or delete items. When PhotoView 360 is added in, the DisplayManager also provides access to PhotoView options.

Document

A file containing a part, assembly, or drawing.

Draft

The degree of taper or angle of a face, usually applied to molds or castings.

Drawing

A 2D representation of a 3D part or assembly. The extension for a SOLIDWORKS drawing file name is .SLDDRW.

Drawing sheet
A page in a drawing document.

Driven dimension
Measurements of the model, but they do not drive the model and their values cannot be changed.

Driving dimension
Also referred to as a model dimension, it sets the value for a sketch entity. It can also control distance, thickness, and feature parameters.

Edge
A single outside boundary of a feature.

Edge flange
A sheet metal feature that combines a bend and a tab in a single operation.

Equation
Creates a mathematical relation between sketch dimensions, using dimension names as variables, or between feature parameters, such as the depth of an extruded feature or the instance count in a pattern.

Exploded view
Shows an assembly with its components separated from one another, usually to show how to assemble the mechanism.

Export
Save a SOLIDWORKS document in another format for use in other CAD/CAM, rapid prototyping, web, or graphics software applications.

Extension line
The line extending from the model indicating the point from which a dimension is measured.

Extrude
A feature that linearly projects a sketch to either add material to a part (in a base or boss) or remove material from a part (in a cut or hole).

Face
A selectable area (planar or otherwise) of a model or surface with boundaries that help define the shape of the model or surface. For example, a rectangular solid has six faces.

Fasteners
A SOLIDWORKS Toolbox library that adds fasteners automatically to holes in an assembly.

Feature

An individual shape that, combined with other features, makes up a part or assembly. Some features, such as bosses and cuts, originate as sketches. Other features, such as shells and fillets, modify a feature's geometry. However, not all features have associated geometry. Features are always listed in the FeatureManager design tree.

FeatureManager design tree

Located on the left side of the SOLIDWORKS window, it provides an outline view of the active part, assembly, or drawing.

Fill

A solid area hatch or crosshatch. Fill also applies to patches on surfaces.

Fillet

An internal rounding of a corner or edge in a sketch, or an edge on a surface or solid.

Forming tool

Dies that bend, stretch, or otherwise form sheet metal to create such form features as louvers, lances, flanges, and ribs.

Fully defined

A sketch where all lines and curves in the sketch, and their positions, are described by dimensions or relations, or both, and cannot be moved. Fully defined sketch entities are shown in black.

Geometric tolerance

A set of standard symbols that specify the geometric characteristics and dimensional requirements of a feature.

Graphics area

The area in the SOLIDWORKS window where the part, assembly, or drawing appears.

Guide curve

A 2D or 3D curve used to guide a sweep or loft.

Handle

An arrow, square, or circle that you can drag to adjust the size or position of an entity (a feature, dimension, or sketch entity, for example).

Helix

A curve defined by pitch, revolutions, and height. A helix can be used, for example, as a path for a swept feature cutting threads in a bolt.

Hem

A sheet metal feature that folds back at the edge of a part. A hem can be open, closed, double, or tear-drop.

HLR

(Hidden lines removed) a view mode in which all edges of the model that are not visible from the current view angle are removed from the display.

HLV

(Hidden lines visible) A view mode in which all edges of the model that are not visible from the current view angle are shown gray or dashed.

Import

Open files from other CAD software applications into a SOLIDWORKS document.

In-context feature

A feature with an external reference to the geometry of another component; the in-context feature changes automatically if the geometry of the referenced model or feature changes.

• Inference

The system automatically creates (infers) relations between dragged entities (sketched entities, annotations, and components) and other entities and geometry. This is useful when positioning entities relative to one another.

Instance

An item in a pattern or a component in an assembly that occurs more than once. Blocks are inserted into drawings as instances of block definitions.

Interference detection

A tool that displays any interference between selected components in an assembly.

Jog

A sheet metal feature that adds material to a part by creating two bends from a sketched line.

Knit

A tool that combines two or more faces or surfaces into one. The edges of the surfaces must be adjacent and not overlapping, but they cannot ever be planar. There is no difference in the appearance of the face or the surface after knitting.

Layout sketch

A sketch that contains important sketch entities, dimensions, and relations. You reference the entities in the layout sketch when creating new sketches, building new geometry, or positioning components in an assembly. This allows for easier updating of your model because changes you make to the layout sketch propagate to the entire model.

Leader

A solid line from an annotation (note, dimension, and so on) to the referenced feature.

Library feature

A frequently used feature, or combination of features, that is created once and then saved for future use.

Lightweight

A part in an assembly or a drawing has only a subset of its model data loaded into memory. The remaining model data is loaded on an as-needed basis. This improves performance of large and complex assemblies.

Line

A straight sketch entity with two endpoints. A line can be created by projecting an external entity such as an edge, plane, axis, or sketch curve into the sketch.

Loft

A base, boss, cut, or surface feature created by transitions between profiles.

Lofted bend

A sheet metal feature that produces a roll form or a transitional shape from two open profile sketches. Lofted bends often create funnels and chutes.

Mass properties

A tool that evaluates the characteristics of a part or an assembly such as volume, surface area, centroid, and so on.

Mate

A geometric relationship, such as coincident, perpendicular, tangent, and so on, between parts in an assembly.

Mate reference

Specifies one or more entities of a component to use for automatic mating. When you drag a component with a mate reference into an assembly, the software tries to find other combinations of the same mate reference name and mate type.

Mates folder

A collection of mates that are solved together. The order in which the mates appear within the Mates folder does not matter.

Mirror

(a) A mirror feature is a copy of a selected feature, mirrored about a plane or planar face.
(b) A mirror sketch entity is a copy of a selected sketch entity that is mirrored about a centerline.

Miter flange

A sheet metal feature that joins multiple edge flanges together and miters the corner.

Model

3D solid geometry in a part or assembly document. If a part or assembly document contains multiple configurations, each configuration is a separate model.

Model dimension

A dimension specified in a sketch or a feature in a part or assembly document that defines some entity in a 3D model.

Model item

A characteristic or dimension of feature geometry that can be used in detailing drawings.

Model view

A drawing view of a part or assembly.

Mold

A set of manufacturing tooling used to shape molten plastic or other material into a designed part. You design the mold using a sequence of integrated tools that result in cavity and core blocks that are derived parts of the part to be molded.

Motion Study

Motion Studies are graphical simulations of motion and visual properties with assembly models. Analogous to a configuration, they do not actually change the original assembly model or its properties. They display the model as it changes based on simulation elements you add.

Multibody part

A part with separate solid bodies within the same part document. Unlike the components in an assembly, multibody parts are not dynamic.

Native format

DXF and DWG files remain in their original format (are not converted into SOLIDWORKS format) when viewed in SOLIDWORKS drawing sheets (view only).

Open profile

Also called an open contour, it is a sketch or sketch entity with endpoints exposed. For example, a U-shaped profile is open.

Ordinate dimensions

A chain of dimensions measured from a zero ordinate in a drawing or sketch.

Origin

The model origin appears as three gray arrows and represents the (0,0,0) coordinate of the model. When a sketch is active, a sketch origin appears in red and represents the (0,0,0) coordinate of the sketch. Dimensions and relations can be added to the model origin, but not to a sketch origin.

Out-of-context feature

A feature with an external reference to the geometry of another component that is not open.

Over defined

A sketch is over defined when dimensions or relations are either in conflict or redundant.

Parameter

A value used to define a sketch or feature (often a dimension).

Parent

An existing feature upon which other features depend. For example, in a block with a hole, the block is the parent to the child hole feature.

Part

A single 3D object made up of features. A part can become a component in an assembly, and it can be represented in 2D in a drawing. Examples of parts are bolt, pin, plate, and so on. The extension for a SOLIDWORKS part file name is .SLDPRT.

Path

A sketch, edge, or curve used in creating a sweep or loft.

Pattern

A pattern repeats selected sketch entities, features, or components in an array, which can be linear, circular, or sketch-driven. If the seed entity is changed, the other instances in the pattern update.

Physical Dynamics

An assembly tool that displays the motion of assembly components in a realistic way. When you drag a component, the component applies a force to other components it touches. Components move only within their degrees of freedom.

Pierce relation

Makes a sketch point coincident to the location at which an axis, edge, line, or spline pierces the sketch plane.

Planar

Entities that can lie on one plane. For example, a circle is planar, but a helix is not.

Plane

Flat construction geometry. Planes can be used for a 2D sketch, section view of a model, a neutral plane in a draft feature, and others.

Point

A singular location in a sketch, or a projection into a sketch at a single location of an external entity (origin, vertex, axis, or point in an external sketch).

Predefined view

A drawing view in which the view position, orientation, and so on can be specified before a model is inserted. You can save drawing documents with predefined views as templates.

Profile

A sketch entity used to create a feature (such as a loft) or a drawing view (such as a detail view). A profile can be open (such as a U shape or open spline) or closed (such as a circle or closed spline).

Projected dimension

If you dimension entities in an isometric view, projected dimensions are the flat dimensions in 2D.

Projected view

A drawing view projected orthogonally from an existing view.

PropertyManager

Located on the left side of the SOLIDWORKS window, it is used for dynamic editing of sketch entities and most features.

RealView graphics

A hardware (graphics card) support of advanced shading in real time; the rendering applies to the model and is retained as you move or rotate a part.

Rebuild

Tool that updates (or regenerates) the document with any changes made since the last time the model was rebuilt. Rebuild is typically used after changing a model dimension.

Reference dimension

A dimension in a drawing that shows the measurement of an item, but cannot drive the model and its value cannot be modified. When model dimensions change, reference dimensions update.

Reference geometry

Includes planes, axes, coordinate systems, and 3D curves. Reference geometry is used to assist in creating features such as lofts, sweeps, drafts, chamfers, and patterns.

Relation

A geometric constraint between sketch entities or between a sketch entity and a plane, axis, edge, or vertex. Relations can be added automatically or manually.

Relative view

A relative (or relative to model) drawing view is created relative to planar surfaces in a part or assembly.

Reload

Refreshes shared documents. For example, if you open a part file for read-only access while another user makes changes to the same part, you can reload the new version, including the changes.

Reorder

Reordering (changing the order of) items is possible in the FeatureManager design tree. In parts, you can change the order in which features are solved. In assemblies, you can control the order in which components appear in a bill of materials.

Replace

Substitutes one or more open instances of a component in an assembly with a different component.

Resolved

A state of an assembly component (in an assembly or drawing document) in which it is fully loaded in memory. All the component's model data is available, so its entities can be selected, referenced, edited, and used in mates, and so on.

Revolve

A feature that creates a base or boss, a revolved cut, or revolved surface by revolving one or more sketched profiles around a centerline.

Rip

A sheet metal feature that removes material at an edge to allow a bend.

Rollback

Suppresses all items below the rollback bar.

Section

Another term for profile in sweeps.

Section line
A line or centerline sketched in a drawing view to create a section view.

Section scope
Specifies the components to be left uncut when you create an assembly drawing section view.

Section view
A section view (or section cut) is (1) a part or assembly view cut by a plane, or (2) a drawing view created by cutting another drawing view with a section line.

Seed
A sketch or an entity (a feature, face, or body) that is the basis for a pattern. If you edit the seed, the other entities in the pattern are updated.

Shaded
Displays a model as a colored solid.

Shared values
Also called linked values, these are named variables that you assign to set the value of two or more dimensions to be equal.

Sheet format
Includes page size and orientation, standard text, borders, title blocks, and so on. Sheet formats can be customized and saved for future use. Each sheet of a drawing document can have a different format.

Shell
A feature that hollows out a part, leaving open the selected faces and thin walls on the remaining faces. A hollow part is created when no faces are selected to be open.

Sketch
A collection of lines and other 2D objects on a plane or face that forms the basis for a feature such as a base or a boss. A 3D sketch is non-planar and can be used to guide a sweep or loft, for example.

Smart Fasteners
Automatically adds fasteners (bolts and screws) to an assembly using the SOLIDWORKS Toolbox library of fasteners.

SmartMates
An assembly mating relation that is created automatically.

Solid sweep

A cut sweep created by moving a tool body along a path to cut out 3D material from a model.

Spiral

A flat or 2D helix, defined by a circle, pitch, and number of revolutions.

Spline

A sketched 2D or 3D curve defined by a set of control points.

Split line

Projects a sketched curve onto a selected model face, dividing the face into multiple faces so that each can be selected individually. A split line can be used to create draft features, to create face blend fillets, and to radiate surfaces to cut molds.

Stacked balloon

A set of balloons with only one leader. The balloons can be stacked vertically (up or down) or horizontally (left or right).

Standard 3 views

The three orthographic views (front, right, and top) that are often the basis of a drawing.

StereoLithography

The process of creating rapid prototype parts using a faceted mesh representation in STL files.

Sub-assembly

An assembly document that is part of a larger assembly. For example, the steering mechanism of a car is a sub-assembly of the car.

Suppress

Removes an entity from the display and from any calculations in which it is involved. You can suppress features, assembly components, and so on. Suppressing an entity does not delete the entity; you can unsuppress the entity to restore it.

Surface

A zero-thickness planar or 3D entity with edge boundaries. Surfaces are often used to create solid features. Reference surfaces can be used to modify solid features.

Sweep

Creates a base, boss, cut, or surface feature by moving a profile (section) along a path. For cut-sweeps, you can create solid sweeps by moving a tool body along a path.

Tangent arc

An arc that is tangent to another entity, such as a line.

Tangent edge
The transition edge between rounded or filleted faces in hidden lines visible or hidden lines removed modes in drawings.

Task Pane
Located on the right-side of the SOLIDWORKS window, the Task Pane contains SOLIDWORKS Resources, the Design Library, and the File Explorer.

Template
A document (part, assembly, or drawing) that forms the basis of a new document. It can include user-defined parameters, annotations, predefined views, geometry, and so on.

Temporary axis
An axis created implicitly for every conical or cylindrical face in a model.

Thin feature
An extruded or revolved feature with constant wall thickness. Sheet metal parts are typically created from thin features.

TolAnalyst
A tolerance analysis application that determines the effects that dimensions and tolerances have on parts and assemblies.

Top-down design
An assembly modeling technique where you create parts in the context of an assembly by referencing the geometry of other components. Changes to the referenced components propagate to the parts that you create in context.

Triad
Three axes with arrows defining the X, Y, and Z directions. A reference triad appears in part and assembly documents to assist in orienting the viewing of models. Triads also assist when moving or rotating components in assemblies.

Under defined
A sketch is under defined when there are not enough dimensions and relations to prevent entities from moving or changing size.

Vertex
A point at which two or more lines or edges intersect. Vertices can be selected for sketching, dimensioning, and many other operations.

Viewports
Windows that display views of models. You can specify one, two, or four viewports. Viewports with orthogonal views can be linked, which links orientation and rotation.

Virtual sharp

A sketch point at the intersection of two entities after the intersection itself has been removed by a feature such as a fillet or chamfer. Dimensions and relations to the virtual sharp are retained even though the actual intersection no longer exists.

Weldment

A multibody part with structural members.

Weldment cut list

A table that tabulates the bodies in a weldment along with descriptions and lengths.

Wireframe

A view mode in which all edges of the part or assembly are displayed.

Zebra stripes

Simulate the reflection of long strips of light on a very shiny surface. They allow you to see small changes in a surface that may be hard to see with a standard display.

Zoom

To simulate movement toward or away from a part or an assembly.

Learn Mold-Tooling Designs with SOLIDWORKS Advanced Techniques Textbook

Index

SOLIDWORKS Quick Guide

 Creates a new document.

 Opens an existing document.

 Saves an active document.

 Make Drawing from Part/Assembly.

 Make Assembly from Part/Assembly.

 Prints the active document.

 Print preview.

 Cuts the selection & puts it on the clipboard.

 Copies the selection & puts it on the clipboard.

 Inserts the clipboard contents.

 Deletes the selection.

 Reverses the last action.

 Rebuilds the part / assembly / drawing.

 Redo the last action that was undone.

 Saves all documents.

 Edits material.

 Closes an existing document.

 Shows or hides the Selection Filter toolbar.

 Shows or hides the Web toolbar.

 Properties.

 File properties.

 Loads or unloads the 3D instant website add-in.

 Select tool.

 Select the entire document.

 Checks read-only files.

 Options.

 Help.

 Full screen view.

 OK.

 Cancel.

 Magnified selection.

 Select.

 Sketch.

 3D Sketch.

 Sketches a rectangle from the center.

 Sketches a centerpoint arc slot.

 Sketches a 3-point arc slot.

 Sketches a straight slot.

 Sketches a centerpoint straight slot.

 Sketches a 3-point arc.

 Creates sketched ellipses.

SKETCH TOOLS Toolbar

3D sketch on plane.

Sets up Grid parameters.

Creates a sketch on a selected plane or face.

Equation driven curve

Modifies a sketch.

Copies sketch entities.

Scales sketch entities.

Rotates sketch entities.

Sketches 3 point rectangle from the center.

Sketches 3 point corner rectangle.

Sketches a line.

Creates a center point arc: center, start, end.

Creates an arc tangent to a line.

Sketches splines on a surface or face.

Sketches a circle.

Sketches a circle by its perimeter.

Makes a path of sketch entities.

Mirrors entities dynamically about a centerline.

Insert a plane into the 3D sketch.

Instant 2D.

Sketch numeric input.

Detaches segment on drag.

Sketch picture.

Partial ellipses.

Adds a Parabola.

Adds a spline.

Sketches a polygon.

Sketches a corner rectangle.

Sketches a parallelogram.

Creates points.

Creates sketched centerlines.

Adds text to sketch.

Converts selected model edges or sketch entities to sketch segments.

Creates a sketch along the intersection of multiple bodies.

Converts face curves on the selected face into 3D sketch entities.

Mirrors selected segments about a centerline.

Fillets the corner of two lines.

Creates a chamfer between two sketch entities.

Creates a sketch curve by offsetting model edges or sketch entities at a specified distance.

Trims a sketch segment.

Extends a sketch segment.

Splits a sketch segment.

Construction Geometry.

Creates linear steps and repeat of sketch entities.

Creates circular steps and repeat of sketch entities.

SHEET METAL Toolbar

 Add a bend from a selected sketch in a Sheet Metal part.

 Shows flat pattern for this sheet metal part.

 Shows part without inserting any bends.

 Inserts a rip feature to a sheet metal part.

 Create a Sheet Metal part or add material to existing Sheet Metal part.

 Inserts a Sheet Metal Miter Flange feature.

 Folds selected bends.

 Unfolds selected bends.

 Inserts bends using a sketch line.

 Inserts a flange by pulling an edge.

 Inserts a sheet metal corner feature.

 Inserts a Hem feature by selecting edges.

 Breaks a corner by filleting/chamfering it.

 Inserts a Jog feature using a sketch line.

 Inserts a lofted bend feature using 2 sketches.

 Creates inverse dent on a sheet metal part.

 Trims out material from a corner, in a sheet metal part.

 Inserts a fillet weld bead.

 Converts a solid/surface into a sheet metal part.

 Adds a Cross Break feature into a selected face.

 Sweeps an open profile along an open/closed path.

 Adds a gusset/rib across a bend.

 Corner relief.

 Welds the selected corner.

SURFACES Toolbar

 Creates mid surfaces between offset face pairs.

 Patches surface holes and external edges.

 Creates an extruded surface.

 Creates a revolved surface.

 Creates a swept surface.

 Creates a lofted surface.

 Creates an offset surface.

 Radiates a surface originating from a curve, parallel to a plane.

 Knits surfaces together.

 Creates a planar surface from a sketch or a set of edges.

 Creates a surface by importing data from a file.

 Extends a surface.

 Trims a surface.

 Surface flatten.

 Deletes Face(s).

 Replaces Face with Surface.

 Patches surface holes and external edges by extending the surfaces.

 Creates parting surfaces between core & cavity surfaces.

 Inserts ruled surfaces from edges.

WELDMENTS Toolbar

 Creates a weldment feature.

 Creates a structure member feature.

 Adds a gusset feature between 2 planar adjoining faces.

 Creates an end cap feature.

 Adds a fillet weld bead feature.

 Trims or extends structure members.

 Weld bead.

DIMENSIONS/RELATIONS Toolbar

 Inserts dimension between two lines.

 Creates a horizontal dimension between selected entities.

 Creates a vertical dimension between selected entities.

 Creates a reference dimension between selected entities.

 Creates a set of ordinate dimensions.

 Creates a set of Horizontal ordinate dimensions.

 Creates a set of Vertical ordinate dimensions.

 Creates a chamfer dimension.

 Adds a geometric relation.

 Automatically Adds Dimensions to the current sketch.

 Displays and deletes geometric relations.

 Fully defines a sketch.

 Scans a sketch for elements of equal length or radius.

 Angular Running dimension.

 Display / Delete dimension.

 Isolate changed dimension.

 Path length dimension.

BLOCK Toolbar

 Makes a new block.

 Edits the selected block.

 Inserts a new block to a sketch or drawing.

 Adds/Removes sketch entities to/from blocks.

 Updates parent sketches affected by this block.

 Saves the block to a file.

 Explodes the selected block.

 Inserts a belt.

STANDARD VIEWS Toolbar

 Front view.

 Back view.

 Left view.

 Right view.

 Top view.

 Bottom view.

 Isometric view.

 Trimetric view.

 Dimetric view.

 Normal to view.

 Links all views in the viewport together.

 Displays viewport with front & right views.

 Displays a 4 view viewport with 1st or 3rd angle of projection.

 Displays viewport with front & top views.

 Displays viewport with a single view.

 View selector.

 New view.

FEATURES Toolbar

 Creates a boss feature by extruding a sketched profile.

 Creates a revolved feature based on profile and angle parameter.

 Creates a cut feature by extruding a sketched profile.

 Creates a cut feature by revolving a sketched profile.

 Thread.

 Creates a cut by sweeping a closed profile along an open or closed path.

 Loft cut.

 Creates a cut by thickening one or more adjacent surfaces.

 Adds a deformed surface by push or pull on points.

 Creates a lofted feature between two or more profiles.

 Creates a solid feature by thickening one or more adjacent surfaces.

 Creates a filled feature.

 Chamfers an edge or a chain of tangent edges.

 Inserts a rib feature.

 Combine.

 Creates a shell feature.

 Applies draft to a selected surface.

 Creates a cylindrical hole.

 Inserts a hole with a pre-defined cross section.

 Puts a dome surface on a face.

 Model break view.

 Applies global deformation to solid or surface bodies.

 Wraps closed sketch contour(s) onto a face.

 Curve Driven pattern.

 Suppresses the selected feature or component.

 Un-suppresses the selected feature or component.

 Flexes solid and surface bodies.

 Intersect.

 Variable Patterns.

 Live Section Plane.

 Mirrors.

 Scale.

 Creates a Sketch Driven pattern.

 Creates a Table Driven Pattern.

 Inserts a split Feature.

 Hole series.

 Joins bodies from one or more parts into a single part in the context of an assembly.

 Deletes a solid or a surface.

 Instant 3D.

 Inserts a part from file into the active part document.

 Moves/Copies solid and surface bodies or moves graphics bodies.

 Merges short edges on faces.

 Pushes solid / surface model by another solid / surface model.

 Moves face(s) of a solid.

 FeatureWorks Options.

 Linear Pattern.

 Fill Pattern.

 Cuts a solid model with a surface.

 Boundary Boss/Base.

 Boundary Cut.

 Circular Pattern.

 Recognize Features.

 Grid System.

MOLD TOOLS Toolbar

 Extracts core(s) from existing tooling split.

 Constructs a surface patch.

 Moves face(s) of a solid.

 Creates offset surfaces.

 Inserts cavity into a base part.

 Scales a model by a specified factor.

 Applies draft to a selected surface.

 Inserts a split line feature.

 Creates parting lines to separate core & cavity surfaces.

 Finds & creates mold shut-off surfaces.

 Creates a planar surface from a sketch or a set of edges.

 Knits surfaces together.

 Inserts ruled surfaces from edges.

 Creates parting surfaces between core & cavity surfaces.

 Creates multiple bodies from a single body.

 Inserts a tooling split feature.

 Creates parting surfaces between the core & cavity.

 Inserts surface body folders for mold operation.

SELECTION FILTERS Toolbar

 Turns selection filters on and off.

 Clears all filters.

 Selects all filters.

 Inverts current selection.

 Allows selection of edges only.

 Allows selection filter for vertices only.

 Allows selection of faces only.

 Adds filter for Surface Bodies.

 Adds filter for Solid Bodies.

 Adds filter for Axes.

 Adds filter for Planes.

 Adds filter for Sketch Points.

 Allows selection for sketch only.

 Adds filter for Sketch Segments.

 Adds filter for Midpoints.

 Adds filter for Center Marks.

 Adds filter for Centerline.

 Adds filter for Dimensions and Hole Callouts.

 Adds filter for Surface Finish Symbols.

 Adds filter for Geometric Tolerances.

 Adds filter for Notes / Balloons.

 Adds filter for Weld Symbols.

 Adds filter for Weld beads.

 Adds filter for Datum Targets.

 Adds filter for Datum feature only.

 Adds filter for blocks.

 Adds filter for Cosmetic Threads.

Adds filter for Dowel pin symbols.

 Adds filter for connection points.

 Adds filter for routing points.

SOLIDWORKS Add-Ins Toolbar

 Loads/unloads CircuitWorks add-in.

 Loads/unloads the Design Checker add-in.

 Loads/unloads the PhotoView 360 add-in.

 Loads/unloads the Scan-to-3D add-in.

 Loads/unloads the SOLIDWORKS Motions add-in.

 Loads/unloads the SOLIDWORKS Routing add-in.

 Loads/unloads the SOLIDWORKS Simulation add-in.

 Loads/unloads the SOLIDWORKS Toolbox add-in.

 Loads/unloads the SOLIDWORKS TolAnalysis add-in.

 Loads/unloads the SOLIDWORKS Flow Simulation add-in.

 Loads/unloads the SOLIDWORKS Plastics add-in.

 Loads/unloads the SOLIDWORKS MBD SNL license.

FASTENING FEATURES Toolbar

 Creates a parameterized mounting boss.

 Creates a parameterized snap hook.

 Creates a groove to mate with a hook feature.

 Uses sketch elements to create a vent for air flow.

 Creates a lip/groove feature.

SCREEN CAPTURE Toolbar

 Copies the current graphics window to the clipboard.

 Records the current graphics window to an AVI file.

 Stops recording the current graphics window to an AVI file.

EXPLODE LINE SKETCH Toolbar

 Adds a route line that connects entities.

 Adds a jog to the route lines.

LINE FORMAT Toolbar

 Changes layer properties.

 Changes the current document layer.

 Changes line color.

 Changes line thickness.

 Changes line style.

 Hides / Shows a hidden edge.

 Changes line display mode.

 Did you know??

* Ctrl+Q will force a rebuild on all features of a part.

* Ctrl+B will rebuild the feature being worked on and its dependents.

2D-To-3D Toolbar

 Makes a Front sketch from the selected entities.

 Makes a Top sketch from the selected entities.

 Makes a Right sketch from the selected entities.

Makes a Left sketch from the selected entities.

Makes a Bottom sketch from the selected entities.

Makes a Back sketch from the selected entities.

Makes an Auxiliary sketch from the selected entities.

Creates a new sketch from the selected entities.

Repairs the selected sketch.

Aligns a sketch to the selected point.

Creates an extrusion from the selected sketch segments, starting at the selected sketch point.

Creates a cut from the selected sketch segments, optionally starting at the selected sketch point.

ALIGN Toolbar

Aligns the left side of the selected annotations with the leftmost annotation.

Aligns the right side of the selected annotations with the rightmost annotation.

Aligns the top side of the selected annotations with the topmost annotation.

Aligns the bottom side of the selected annotations with the lowermost annotation.

Evenly spaces the selected annotations horizontally.

Evenly spaces the selected annotations vertically.

Centrally aligns the selected annotations horizontally.

Centrally aligns the selected annotations vertically.

Compacts the selected annotations horizontally.

Compacts the selected annotations vertically.

Creates a group from the selected items.

Deletes the grouping between these items.

Aligns & groups selected dimensions along a line or an arc.

Aligns & groups dimensions at uniform distances.

Evenly spaces selected dimensions.

Aligns collinear selected dimensions.

Aligns stagger selected dimensions.

SOLIDWORKS MBD Toolbar

Captures 3D view.

Manages 3D PDF templates.

Creates shareable 3D PDF presentations.

Toggles dynamic annotation views.

MACRO Toolbar

Runs a Macro.

Stops Macro recorder.

Records (or pauses recording of) actions to create a Macro.

Launches the Macro Editor and begins editing a new macro.

Opens a Macro file for editing.

Creates a custom macro.

SMARTMATES icons

Concentric & Coincident 2 circular edges.

Concentric 2 cylindrical faces.

Coincident 2 linear edges.

Coincident 2 planar faces.

Coincident 2 vertices.

Coincident 2 origins or coordinate systems.

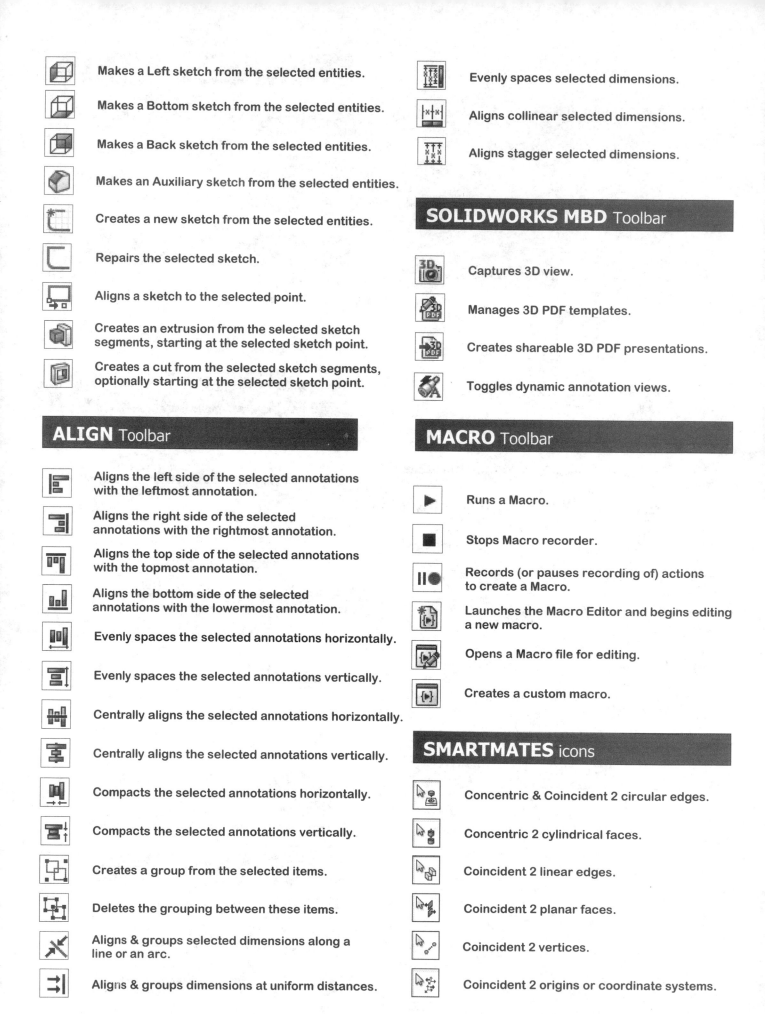

TABLE Toolbar

 Adds a hole table of selected holes from a specified origin datum.

 Adds a Bill of Materials.

 Adds a revision table.

 Displays a Design table in a drawing.

 Adds a weldments cuts list table.

 Adds an Excel based Bill of Materials.

 Adds a weldment cut list table.

REFERENCE GEOMETRY Toolbar

 Adds a reference plane.

 Creates an axis.

 Creates a coordinate system.

 Adds the center of mass.

 Specifies entities to use as references using SmartMates.

SPLINE TOOLS Toolbar

 Inserts a point to a spline.

 Displays all points where the concavity of selected spline changes.

 Displays minimum radius of selected spline.

 Displays curvature combs of selected spline.

 Reduces numbers of points in a selected spline.

 Adds a tangency control.

 Adds a curvature control.

 Adds a spline based on selected sketch entities & edges.

 Displays the spline control polygon.

ANNOTATIONS Toolbar

 Inserts a note.

 Inserts a surface finish symbol.

 Inserts a new geometric tolerancing symbol.

 Attaches a balloon to the selected edge or face.

 Adds balloons for all components in selected view.

 Inserts a stacked balloon.

 Attaches a datum feature symbol to a selected edge / detail.

 Inserts a weld symbol on the selected edge / face / vertex.

 Inserts a datum target symbol and / or point attached to a selected edge / line.

 Selects and inserts block.

 Inserts annotations & reference geometry from the part / assembly into the selected.

 Adds center marks to circles on model.

 Inserts a Centerline.

 Inserts a hole callout.

 Adds a cosmetic thread to the selected cylindrical feature.

 Inserts a Multi-Jog leader.

 Selects a circular edge or an arc for Dowel pin symbol insertion.

 Adds a view location symbol.

 Inserts latest version symbol.

 Adds cross hatch patterns or solid fill.

 Adds a weld bead caterpillar on an edge.

 Adds a weld symbol on a selected entity.

 Inserts a revision cloud.

 Inserts a magnetic line.

 Hides/shows annotation.

DRAWINGS Toolbar

 Updates the selected view to the model's current stage.

 Creates a detail view.

 Creates a section view.

 Inserts an Alternate Position view.

 Unfolds a new view from an existing view.

 Generates a standard 3-view drawing (1st or 3rd angle).

 Inserts an auxiliary view of an inclined surface.

 Adds an Orthogonal or Named view based on an existing part or assembly.

 Adds a Relative view by two orthogonal faces or planes.

 Adds a Predefined orthogonal projected or Named view with a model.

 Adds an empty view.

 Adds vertical break lines to selected view.

 Crops a view.

Creates a Broken-out section.

QUICK SNAP Toolbar

 Snap to points.

 Snap to center points.

 Snap to midpoints.

 Snap to quadrant points.

 Snap to intersection of 2 curves.

 Snap to nearest curve.

 Snap tangent to curve.

 Snap perpendicular to curve.

 Snap parallel to line.

 Snap horizontally / vertically to points.

 Snap horizontally / vertically.

 Snap to discrete line lengths.

 Snap to angle.

LAYOUT Toolbar

 Creates the assembly layout sketch.

 Sketches a line.

 Sketches a corner rectangle.

 Sketches a circle.

 Sketches a 3 point arc.

 Rounds a corner.

 Trims or extends a sketch.

 Adds sketch entities by offsetting faces, edges and curves.

 Mirrors selected entities about a centerline.

 Adds a relation.

 Creates a dimension.

 Displays / Deletes geometric relations.

 Makes a new block.

 Edits the selected block.

 Inserts a new block to the sketch or drawing.

 Adds / Removes sketch entities to / from a block.

 Saves the block to a file.

 Explodes the selected block.

 Creates a new part from a layout sketch block.

 Positions 2 components relative to one another.

CURVES Toolbar

 Projects sketch onto selected surface.

 Inserts a split line feature.

 Creates a composite curve from selected edges, curves and sketches.

 Creates a curve through free points.

 Creates a 3D curve through reference points.

 Helical curve defined by a base sketch and shape parameters.

VIEW Toolbar

 Displays a view in the selected orientation.

 Reverts to previous view.

 Redraws the current window.

 Zooms out to see entire model.

 Zooms in by dragging a bounding box.

 Zooms in or out by dragging up or down.

 Zooms to fit all selected entities.

 Dynamic view rotation.

 Scrolls view by dragging.

 Displays image in wireframe mode.

 Displays hidden edges in gray.

 Displays image with hidden lines removed.

 Controls the visibility of planes.

 Controls the visibility of axis.

 Controls the visibility of parting lines.

 Controls the visibility of temporary axis.

 Controls the visibility of origins.

 Controls the visibility of coordinate systems.

 Controls the visibility of reference curves.

 Controls the visibility of sketches.

 Controls the visibility of 3D sketch planes.

 Controls the visibility of 3D sketch.

 Controls the visibility of all annotations.

 Controls the visibility of reference points.

 Controls the visibility of routing points.

 Controls the visibility of lights.

 Controls the visibility of cameras.

 Controls the visibility of sketch relations.

 Changes the display state for the current configuration.

 Rolls the model view.

 Turns the orientation of the model view.

 Dynamically manipulate the model view in 3D to make selection.

 Changes the display style for the active view.

 Displays a shade view of the model with its edges.

 Displays a shade view of the model.

 Toggles between draft quality & high quality HLV.

 Cycles through or applies a specific scene.

 Views the models through one of the model's cameras.

 Displays a part or assembly w/different colors according to the local radius of curvature.

 Displays zebra stripes.

 Displays a model with hardware accelerated shades.

 Applies a cartoon effect to model edges & faces.

 Views simulations symbols.

TOOLS Toolbar

 Calculates the distance between selected items.

 Adds or edits equation.

 Calculates the mass properties of the model.

 Checks the model for geometry errors.

 Inserts or edits a Design Table.

 Evaluates section properties for faces and sketches that lie in parallel planes.

 Reports Statistics for this Part/Assembly.

 Deviation Analysis.

 Runs the SimulationXpress analysis wizard Powered by SOLIDWORKS Simulation.

 Checks the spelling.

 Import diagnostics.

 Runs the DFMXpress analysis wizard.

 Runs the SOLIDWORKSFloXpress analysis wizard.

ASSEMBLY Toolbar

 Creates a new part & inserts it into the assembly.

 Adds an existing part or sub-assembly to the assembly.

 Creates a new assembly & inserts it into the assembly.

 Turns on/off large assembly mode for this document.

 Hides / shows model(s) associated with the selected model(s).

 Toggles the transparency of components.

 Changes the selected components to suppressed or resolved.

 Inserts a belt.

 Toggles between editing part and assembly.

 Smart Fasteners.

 Positions two components relative to one another.

 External references will not be created.

 Moves a component.

 Rotates an un-mated component around its center point.

 Replaces selected components.

 Replaces mate entities of mates of the selected components on the selected Mates group.

 Creates a New Exploded view.

 Creates or edits explode line sketch.

 Interference detection.

 Shows or Hides the Simulation toolbar.

 Patterns components in one or two linear directions.

 Patterns components around an axis.

 Sets the transparency of the components other than the one being edited.

 Sketch driven component pattern.

 Pattern driven component pattern.

 Curve driven component pattern.

 Chain driven component pattern.

 SmartMates by dragging & dropping components.

 Checks assembly hole alignments.

 Mirrors subassemblies and parts.

To add or remove an icon
to or from the toolbar, first select:

Tools/Customize/Commands

Next, select a **Category**, click a button to see its description and then drag/drop the command icon into any toolbar.

SOLIDWORKS Quick-Guide©
STANDARD Keyboard Shortcuts

Rotate the model

* Horizontally or Vertically: _____ Arrow keys

* Horizontally or Vertically 90°: _____ Shift + Arrow keys

* Clockwise or Counterclockwise: _____ Alt + left or right Arrow

* Pan the model: _____ Ctrl + Arrow keys

* Zoom in: _____ Z (shift + Z or capital Z)

* Zoom out: _____ z (lower case z)

* Zoom to fit: _____ F

* Previous view: _____ Ctrl+Shift+Z

View Orientation

* View Orientation Menu: _____ Space bar

* Front: _____ Ctrl+1

* Back: _____ Ctrl+2

* Left: _____ Ctrl+3

* Right: _____ Ctrl+4

* Top: _____ Ctrl+5

* Bottom: _____ Ctrl+6

* Isometric: _____ Ctrl+7

Selection Filter & Misc.

* Filter Edges: _____ e

* Filter Vertices: _____ v

* Filter Faces: _____ x

* Toggle Selection filter toolbar: _____ F5

* Toggle Selection Filter toolbar (on/off): _____ F6

* New SOLIDWORKS document: _____ F1

* Open Document: _____ Ctrl+O

* Open from Web folder: _____ Ctrl+W

* Save: _____ Ctrl+S

* Print: _____ Ctrl+P

* Magnifying Glass Zoom _____ g

* Switch between the SOLIDWORKS documents _____ Ctrl + Tab

SOLIDWORKS Sample Customized Hot Keys

Function Keys

F1	SW-Help
F2	2D Sketch
F3	3D Sketch
F4	Modify
F5	Selection Filters
F6	Move (2D Sketch)
F7	Rotate (2D Sketch)
F8	Measure
F9	Extrude
F10	Revolve
F11	Sweep
F12	Loft

Sketch

C	Circle
P	Polygon
E	Ellipse
O	Offset Entities
Alt + C	Convert Entities
M	Mirror
Alt + M	Dynamic Mirror
Alt + F	Sketch Fillet
T	Trim
Alt + X	Extend
D	Smart Dimension
Alt + R	Add Relation
Alt + P	Plane
Control + F	Fully Define Sketch
Control + Q	Exit Sketch